Grinding and Concentration Technology of Critical Metals

Grinding and Concentration Technology of Critical Metals

Editor

Juan M Menéndez-Aguado

MDPI • Basel • Beijing • Wuhan • Barcelona • Belgrade • Manchester • Tokyo • Cluj • Tianjin

Editor
Juan M Menéndez-Aguado
Polytechnic School of Mieres
University of Oviedo
Mieres
Spain

Editorial Office
MDPI
St. Alban-Anlage 66
4052 Basel, Switzerland

This is a reprint of articles from the Special Issue published online in the open access journal *Metals* (ISSN 2075-4701) (available at: www.mdpi.com/journal/metals/special_issues/grinding_technology).

For citation purposes, cite each article independently as indicated on the article page online and as indicated below:

LastName, A.A.; LastName, B.B.; LastName, C.C. Article Title. *Journal Name* **Year**, *Volume Number*, Page Range.

ISBN 978-3-0365-3956-0 (Hbk)
ISBN 978-3-0365-3955-3 (PDF)

© 2022 by the authors. Articles in this book are Open Access and distributed under the Creative Commons Attribution (CC BY) license, which allows users to download, copy and build upon published articles, as long as the author and publisher are properly credited, which ensures maximum dissemination and a wider impact of our publications.

The book as a whole is distributed by MDPI under the terms and conditions of the Creative Commons license CC BY-NC-ND.

Contents

About the Editor vii

Juan M. Menéndez Aguado
Grinding and Concentration Technology of Critical Metals
Reprinted from: *Metals* **2022**, *12*, 585, doi:10.3390/met12040585 1

Vladimir Nikolić, Gloria G. García, Alfredo L. Coello-Velázquez, Juan M. Menéndez-Aguado, Milan Trumić and Maja S. Trumić
A Review of Alternative Procedures to the Bond Ball Mill Standard Grindability Test
Reprinted from: *Metals* **2021**, *11*, 1114, doi:10.3390/met11071114 5

Victor Ciribeni, Regina Bertero, Andrea Tello, Matías Puerta, Enzo Avellá and Matías Paez et al.
Application of the Cumulative Kinetic Model in the Comminution of Critical Metal Ores
Reprinted from: *Metals* **2020**, *10*, 925, doi:10.3390/met10070925 21

Victor Ciribeni, Juan M. Menéndez-Aguado, Regina Bertero, Andrea Tello, Enzo Avellá and Matías Paez et al.
Unveiling the Link between the Third Law of Comminution and the Grinding Kinetics Behaviour of Several Ores
Reprinted from: *Metals* **2021**, *11*, 1079, doi:10.3390/met11071079 33

Jenniree V. Nava, Teresa Llorens and Juan María Menéndez-Aguado
Kinetics of Dry-Batch Grinding in a Laboratory-Scale Ball Mill of Sn–Ta–Nb Minerals from the Penouta Mine (Spain)
Reprinted from: *Metals* **2020**, *10*, 1687, doi:10.3390/met10121687 45

Jennire V. Nava, Alfredo L. Coello-Velázquez and Juan M. Menéndez-Aguado
Grinding Kinetics Study of Tungsten Ore
Reprinted from: *Metals* **2020**, *11*, 71, doi:10.3390/met11010071 65

Gloria G. García, Josep Oliva, Eduard Guasch, Hernán Anticoi, Alfredo L. Coello-Velázquez and Juan M. Menéndez-Aguado
Variability Study of Bond Work Index and Grindability Index on Various Critical Metal Ores
Reprinted from: *Metals* **2021**, *11*, 970, doi:10.3390/met11060970 79

Gloria González García, Alfredo L. Coello-Velázquez, Begoña Fernández Pérez and Juan M. Menéndez-Aguado
Variability of the Ball Mill Bond's Standard Test in a Ta Ore Due to the Lack of Standardization
Reprinted from: *Metals* **2021**, *11*, 1606, doi:10.3390/met11101606 91

Nikolay Kolev, Petar Bodurov, Vassil Genchev, Ben Simpson, Manuel G. Melero and Juan M. Menéndez-Aguado
A Comparative Study of Energy Efficiency in Tumbling Mills with the Use of Relo Grinding Media
Reprinted from: *Metals* **2021**, *11*, 735, doi:10.3390/met11050735 99

Laura Colorado-Arango, Juan M. Menéndez-Aguado and Adriana Osorio-Correa
Particle Size Distribution Models for Metallurgical Coke Grinding Products
Reprinted from: *Metals* **2021**, *11*, 1288, doi:10.3390/met11081288 111

Angel R. Llera, Ana Díaz, Francisco J. Pedrayes, Juan M. Menéndez-Aguado and Manuel G. Melero
Study of Comminution Kinetics in an Electrofragmentation Lab-Scale Device
Reprinted from: *Metals* **2022**, *12*, 494, doi:10.3390/met12030494 . . . . **123**

About the Editor

Juan M Menéndez-Aguado

Full Professor of Mineral Processing Technology at the University of Oviedo. Supervisor of more than 20 PhD Theses. Participant in more than 120 projects financed by private and public institutions (Research, Technology Transfer and Academic). More than 80 scientific publications (+50 in indexed reviews). Visiting Professor in more than 10 Universities in Argentina, Chile, Colombia, Cuba, Ecuador, Perú and Venezuela. Member of the Society of Mining, Metallurgy and Exploration (SME/AIME). Member of the Society of Mining Professors (SOMP) since 2013.

Editorial

Grinding and Concentration Technology of Critical Metals

Juan M. Menéndez Aguado

Escuela Politécnica de Mieres, University of Oviedo, C/Gonzalo Gutierrez Quirós s/n, 33600 Mieres, Spain; maguado@uniovi.es

1. Introduction and Scope

The production and supply of raw materials in a global market are not without risks, and both the recent COVID-19 pandemic and the current one (Russia–Ukrania conflict) raised public awareness about the importance of multiple value chains.

Despite the great inertia characterising the mineral raw materials sector, some steps towards the Industry 4.0 paradigm can be envisaged. Significant challenges to the mining sector are the appropriate process design using the best available technologies; the increase in energy efficiency; the responsible use of water and handling of mining wastes; the social acceptance of the activity; and the digitalisation challenge. More than ten years ago, the European Union elaborated a list of critical raw materials (CRMs), taking the economic and strategic importance for the European economy and the supply risk. Although focused mainly on the energy sector, the USA, Canada, and other countries took recently similar steps.

This Special Issue aims to propose strategies that can help face those challenges, especially in increasing energy efficiency in comminution operations.

2. Contributions

In the first contribution of this Special Issue, Ciribeni et al. [1] proposed a simplified procedure for calculating grinding kinetic parameters, providing a spreadsheet to help work index calculation through simulation using the characterisation performed. They then compared the results with actual Bond ball-mill work index results and validate the proposed methodology.

Another contribution regarding grinding kinetics of a Ta ore was the research objective in Nava et al. [2]. Some variations to classical population balance model methodologies and functional operational correlations were found among the feed size; the specific breakage rate; and the Sn, Ta, and Nb contents. This study was completed with additional experimental tests on this same ore by Nava et al. [3], obtaining a more profound comprehension about the relationship among each kinetic parameter and the operational conditions (mill speed and feed grain size), which permitted the definition of the operation conditions to improve grinding efficiency.

A different and very innovative approach to energy efficiency improvement in grinding is proposed in the paper authored by Kolev et al. [4]. The substitution of steel balls by Relo grinding media (RGM) in tumbling mills is the focus of the research study. RGM are claimed by the producer, the Bulgarian company RELO-B, as a better alternative for balls. RGM were tested at laboratory scale under different conditions and compared with balls equivalent in diameter. Although standard Bond tests were not conclusive, results were promising in terms of grinding efficiency, reaching, for the RGM, the same undersize production as balls with lower circulating load values. Further research is needed to clarify the effect on mineral liberation.

In the paper proposed by García et al. [5], a deep study on the Bond ball-mill and Bond rod-mill standard tests is shown with different ores. The most impacting result of this work is the different results when matching work index values and grindability index

values from the od mill's size range to the ball mill's size range, showing that the parameter which reflects ore grinding properties is the grindability index, which allowed the work index to have additional influences from operational conditions. Furthermore, the authors proposed the Maxson index when referring to the grindability index, based on the historical importance of Walter Maxson and its mentoring role on Fred Bond's initial research stage at Allis Chalmers laboratories.

The sixth paper in this Special Issue, authored by Ciribeni et al. [6], discussed the relationship between the Maxson Index and the kinetic parameter obtained in the grinding kinetics characterisation of several ores following the Cummulative Kinetic Method (CKM). Up to twelve different ores tested under fifteen different conditions proved that these parameters have a strong correlation, which led the authors to propose a rapid methodology of work index determination.

With all the results presented in previous papers, the time to perform a deep revision arrived, and this task was performed in the paper authored by Nikolić et al. [7]. This excellent review paper, which is not exhaustive but very well focused, revised up to twenty-two alternative procedures to work index determination, performing a revealing comparative of the mean square relative error in each case.

A place in this Special Issue was also left to the research work authored by Colorado-Arango et al. [8]. Although the research addressed the grinding of metallurgical coke, the study focused on the influence of the selection among different particle size distribution (PSD) models when predicting grinding products, with importances when performing interpolations to obtain PSD characteristic sizes (d_{80}).

The ninth contribution to this Special Issue, signed by García et al. [9], discusses the variability on the work index when performing the Bond ball-mill standard test due to the lack of definition of several test conditions. An ANOVA test shows the influence of F_{80}, P_{100} and the feed fines percentage (% < P_{100}), highlighting that, with the same ore, the Bond work index values can show significant differences, and its proper interpretation needs additional information further than the sole value result.

Finally, the last contribution to this Special Issue by Llera et al. [10] proposes a kinetic model of the comminution process in a high voltage impulse electrofragmentation device. The authors studied the influence of feed particle size, impulse number, and impulse polarity on the grinding product and the model parameters, evidencing original conclusions that interest this breakthrough comminution technology.

3. Conclusions and Outlook

The papers published in this Special Issue evidenced that increasing energy efficiency is a major challenge that can be faced with a better understanding of traditional approaches, as is the case of Bond's methodology or grinding-kinetics ore characterisation. However, this major challenge must also consider innovative approaches in the state-of-the-art methodologies, as is the case of the RGM use in conventional tumbling mills or focusing on the scaling up of revolutionary technologies, as is the case of electrofragmentation technology in comminution.

Funding: This research received no external funding.

Conflicts of Interest: The author declares no conflict of interest.

References

1. Ciribeni, V.; Bertero, R.; Tello, A.; Puerta, M.; Avellá, E.; Paez, M.; Menéndez Aguado, J. Application of the Cumulative Kinetic Model in the Comminution of Critical Metal Ores. *Metals* **2020**, *10*, 925. [CrossRef]
2. Nava, J.; Llorens, T.; Menéndez-Aguado, J. Kinetics of Dry-Batch Grinding in a Laboratory-Scale Ball Mill of Sn–Ta–Nb Minerals from the Penouta Mine (Spain). *Metals* **2020**, *10*, 1687. [CrossRef]
3. Nava, J.; Coello-Velázquez, A.; Menéndez-Aguado, J. Grinding Kinetics Study of Tungsten Ore. *Metals* **2021**, *11*, 71. [CrossRef]
4. Kolev, N.; Bodurov, P.; Genchev, V.; Simpson, B.; Melero, M.; Menéndez-Aguado, J. A Comparative Study of Energy Efficiency in Tumbling Mills with the Use of Relo Grinding Media. *Metals* **2021**, *11*, 735. [CrossRef]

5. García, G.; Oliva, J.; Guasch, E.; Anticoi, H.; Coello-Velázquez, A.; Menéndez-Aguado, J. Variability Study of Bond Work Index and Grindability Index on Various Critical Metal Ores. *Metals* **2021**, *11*, 970. [CrossRef]
6. Ciribeni, V.; Menéndez-Aguado, J.; Bertero, R.; Tello, A.; Avellá, E.; Paez, M.; Coello-Velázquez, A. Unveiling the Link between the Third Law of Comminution and the Grinding Kinetics Behaviour of Several Ores. *Metals* **2021**, *11*, 1079. [CrossRef]
7. Nikolić, V.; García, G.; Coello-Velázquez, A.; Menéndez-Aguado, J.; Trumić, M.; Trumić, M. A Review of Alternative Procedures to the Bond Ball Mill Standard Grindability Test. *Metals* **2021**, *11*, 1114. [CrossRef]
8. Colorado-Arango, L.; Menéndez-Aguado, J.; Osorio-Correa, A. Particle Size Distribution Models for Metallurgical Coke Grinding Products. *Metals* **2021**, *11*, 1288. [CrossRef]
9. García, G.; Coello-Velázquez, A.; Pérez, B.; Menéndez-Aguado, J. Variability of the Ball Mill Bond's Standard Test in a Ta Ore Due to the Lack of Standardization. *Metals* **2021**, *11*, 1606. [CrossRef]
10. Llera, A.; Díaz, A.; Pedrayes, F.; Menéndez-Aguado, J.; Melero, M. Study of Comminution Kinetics in an Electrofragmentation Lab-Scale Device. *Metals* **2022**, *12*, 494. [CrossRef]

Review

A Review of Alternative Procedures to the Bond Ball Mill Standard Grindability Test

Vladimir Nikolić [1], Gloria G. García [2], Alfredo L. Coello-Velázquez [3], Juan M. Menéndez-Aguado [2,*], Milan Trumić [1] and Maja S. Trumić [1]

[1] Technical Faculty in Bor, University of Belgrade, 19210 Bor, Serbia; vnikolic@tfbor.bg.ac.rs (V.N.); mtrumic@tfbor.bg.ac.rs (M.T.); majatrumic@tfbor.bg.ac.rs (M.S.T.)
[2] Escuela Politécnica de Mieres, University of Oviedo, Gonzalo Gutiérrez Quirós, 33600 Mieres, Spain; gloria.glez.gcia@gmail.com
[3] CETAM, Universidad de Moa Antonio Núñez Jiménez, Moa 83300, Cuba; acoello@ismm.edu.cu
* Correspondence: maguado@uniovi.es; Tel.: +34-985458033

Abstract: Over the years, alternative procedures to the Bond grindability test have been proposed aiming to avoid the need for the standard mill or to reduce and simplify the grinding procedure. Some of them use the standard mill, while others are based on a non-standard mill or computation techniques. Therefore, papers targeting to propose a better alternative claim to improve validity, to reduce test duration, or to propose simpler and faster alternative methods for determining the Bond work index (w_i). In this review paper, a compilation and critical analysis of selected proposals is performed, concluding that some of the short procedures could be useful for control purposes, while the simulation-based procedures could be interesting within a process digitalisation strategy.

Keywords: grindability; comminution; Bond work index

Citation: Nikolić, V.; García, G.G.; Coello-Velázquez, A.L.; Menéndez-Aguado, J.M.; Trumić, M.; Trumić, M.S. A Review of Alternative Procedures to the Bond Ball Mill Standard Grindability Test. *Metals* **2021**, *11*, 1114. https://doi.org/10.3390/met11071114

Academic Editor: Jean François Blais

Received: 30 May 2021
Accepted: 9 July 2021
Published: 12 July 2021

Publisher's Note: MDPI stays neutral with regard to jurisdictional claims in published maps and institutional affiliations.

Copyright: © 2021 by the authors. Licensee MDPI, Basel, Switzerland. This article is an open access article distributed under the terms and conditions of the Creative Commons Attribution (CC BY) license (https://creativecommons.org/licenses/by/4.0/).

1. Introduction

Determining the Bond index using the Fred Bond method [1,2] is considered the state-of-the-art methodology for mill calculations and a critical process parameter in raw materials selection and grinding process control. Although it is usually referred to as a standard test, no ISO (International Organization for Standardization) or ASTM (American Society for Testing and Materials) standard procedure has been established, so the primary reference used worldwide to define the procedure is the original proposal from Bond. Despite this, the knowledge of the Bond standard test is enriched continuously with new research, as is the case of the recently published work by García et al. [3], which presents a deep analysis of the test procedure and evidences the importance of the grindability index (proposing it to be renamed as the Maxson index), or the recent proposal by Nikolić and Trumić [4], which represent a new approach for determination Bond work index on finer samples.

Alternative tests soon arose after Bond's proposal to avoid the need for the standard mill and time-consuming procedure. Therefore, papers dealing with this problem are numerous, aiming to discuss the validity of simpler and quicker methods to determine the Bond work index (w_i). Some of them use the standard mill, while others use a non-standard mill or are based on computation techniques. In this review paper, a compilation and critical analysis of several selected methodologies were performed, based on the practical experience of the laboratories involved in this research.

It is worthy to mention the development of other approaches to grindability evaluation based on impact breakage tests. The drop weight test has proven its validity and scaling-up possibilities under certain conditions [5–7].

There are not many review papers describing alternative methods of ball mill w_i determination. The work of Lvov and Chitalov [8] is probably the most recent one, and it performs a sound analysis of several alternative methodologies. This review includes additional methodologies and considers the analysis of the relative square error and the procedure advantages claimed by the authors of each proposal.

2. Alternative Procedures to the Bond Ball Mill Standard Test

Berry and Bruce [9] introduced the first alternative procedure to the Bond standard test. The procedure is based on determining the grindability of an unknown ore by comparing it to the grindability behaviour of a reference ore. It can be performed in any laboratory ball mill, but it requires a reference sample ore for which w_i is known. In the Berry and Bruce procedure, 2 kg weight samples of the reference and unknown ores with a particle size under 1.651 mm are wet ground in a laboratory ball mill that is 305 mm in diameter, using active power monitoring. According to the Bond Third's Law of comminution (Equation (1); [1]), after performing both grinding tests with the same specific active power energy consumption, Equation (2) can be deduced:

$$W = 10 \cdot w_i \cdot \left[\frac{1}{\sqrt{P_{80}}} - \frac{1}{\sqrt{F_{80}}} \right] \; [kWh/t] \tag{1}$$

$$w_{ir} \cdot \left[\frac{1}{\sqrt{P_r}} - \frac{1}{\sqrt{F_r}} \right] = w_{i,BB} \cdot \left[\frac{1}{\sqrt{P_{80}}} - \frac{1}{\sqrt{F_{80}}} \right] \Rightarrow w_{i,BB} = w_{ir} \cdot \frac{\left[\frac{1}{\sqrt{P_r}} - \frac{1}{\sqrt{F_r}} \right]}{\left[\frac{1}{\sqrt{P_{80}}} - \frac{1}{\sqrt{F_{80}}} \right]}, \; [kWh/t] \tag{2}$$

wherein:
W—Specific power consumption, (kWh/t);
w_{ir}—Bond work index of reference ore, (kWh/t);
P_r—80% passing product particle size, reference ore, (μm);
F_r—80% passing feed particle size, reference ore, (μm);
$w_{i,BB}$—Bond work index estimation of the unknown ore, (kWh/t);
P_{80}—the 80% passing product particle size, unknown ore, (μm);
F_{80}—the 80% passing feed particle size, unknown ore, (μm).

The validity of this procedure depends on the accuracy of stopping the unknown sample grinding test after a specific power consumption is reached (measured with a power-meter) and on the similarity of the particle size distribution (PSD) of the feed samples. Differences in sample densities and PSD affect the density and rheological characteristics of the pulp when performing the test wet way. Moreover, it has been proven that w_i is not a constant value for each ore, so the reference sample value would only be valid within a specific grinding size range [3,10]. The main advantage of this procedure is that it is fast and does not require Bond's standard ball mill, but accurate power measurements are needed, and the use of a reference ore as if w_i had a constant value is also a source of inaccuracy.

Horst and Bassarear [11] gave a procedure based on Berry and Bruce [9], but with a basis on grinding kinetics. In this case, the procedure does not consider the unknown sample feed and grinding product PSD; instead, starting from the reference ore feed PSD, the unknown ore grinding product PSD is calculated by a first-order kinetics equation (Equation (3)):

$$R = R_0 \cdot e^{-k \cdot t} \tag{3}$$

wherein:
R—oversize of the comparative sieve after grinding time t;
R_0—oversize of the comparative sieve at the beginning of grinding $t = t_0 = 0$;
k—first-order kinetics grinding constant;
t—grinding time.

The test can be performed in any laboratory ball mill on a sample with an initial size under 1.651 mm. A reference ore sample weighing 1 kg is ground until the desired mill PSD is obtained. This can be performed with several grinding tests in a row on the same sample, accumulating the grinding time from one test to another; the sample PSD is obtained after each grinding test, and if a finer product is needed, the sample is returned to the mill and ground for more time. Three unknown ore samples weighing 1 kg are ground in the same mill under the same conditions and with different grinding times. The grinding times of these three samples should include the grinding time of the reference ore. The PSD is determined for all grinding products, and a plot t versus lnR can be performed to obtain the grinding rate constants k_i for each grinding size. The unknown ore grinding product can be calculated using the reference ore feed PSD and the grinding time of the reference ore to the desired fineness, and the value k_i is determined (Equation (3)) in the cycle of ore grindability. Based on the PSD calculated in this way, the value of the parameter P_{80} (µm) is determined, and the value of the parameter F_{80} (µm) is taken to be equal to the parameter F_r. The Bond work index is estimated by Equation (4).

$$w_{i,HB} = w_{ir} \cdot \frac{\left[\frac{1}{\sqrt{P_r}} - \frac{1}{\sqrt{F_r}}\right]}{\left[\frac{1}{\sqrt{P_{80}}} - \frac{1}{\sqrt{F_r}}\right]}, \ [kWh/t] \quad (4)$$

Differences in grindability in this process are reflected only through differences in the size of the grinding product P_{80}. The advantages of this procedure are the use of an ordinary laboratory mill with balls and a smaller mass and sample size than the standard Bond test. A small amount of time is needed to perform the test and calculate the PSD. The total execution time of this test is almost no shorter than the standard Bond test. The Berry–Bruce procedure, which is similar to this test, is considerably shorter and should give more reliable results. However, the mean square relative difference reported by the authors between the values of the Bond work index obtained by the standard method and the values obtained by the Berry–Bruce method is 8.25% and 1.72%. for the Horst–Bassarear method. The relative difference achieved by the Horst–Bassarear procedure is surprisingly small, although the PSD of the feed sample is equated to the PSD of the reference ore feed sample, and the PSD of the unknown ore grinding product is calculated using the grinding kinetics equation.

Smith and Lee [12] determined the Bond work index in a standard mill for eight different materials at different openings of a comparative sieve according to the standard Bond test. They compared the data obtained by the standard Bond test and the data from the open-circuit grinding, i.e., the first grinding cycle of the standard Bond test. The tests showed that the parameter G_z [g/rev] of the last grinding cycle of the standard Bond test and the parameter G_0 of the open-circuit grinding under the same conditions are in a direct correlation $G_0 = f(G_z)$. This correlation was established on screens with smaller openings and in tests performed with less than 300 mill revolutions. With this correlation, it is possible to estimate G_z in the standard Bond test based on the value of G_0 determined in the open circuit grinding, and the estimated Bond work index ($w_{i,SL}$) can then be calculated according to Equation (5).

$$w_{i,SL} = 1.1 \cdot \frac{16}{G^{0.82}} \cdot \sqrt{\frac{P_{100}}{100}} \ [kWh/t] \quad (5)$$

A correlation that is established in this way is valid only for the materials on which it is determined. For other materials, it is necessary to establish a new correlation relationship, which requires a Bond mill and sample preparation conducted in the same way as the standard Bond test. A lot of work and grinding cycles are needed to determine the correlation $G_z = K \cdot G_0$. The Bond work index is estimated based on one grinding cycle performed in the standard mill and calculated by following Equation (5). The Smith–Lee results showed that the differences from the standard Bond test and the $w_{i,SL}$ values do not exceed 15%. Probably one of the main shortcomings of this methodology is the influence of feed particle size on the initial cycles, which could be the main source of deviation.

Kapur [13] analysed the grinding cycles that made up the standard Bond test using a mathematical algorithm based on first order grinding kinetics and concluded that the estimation of w_i could be performed based on the results of the first two grinding cycles from the standard Bond test. In several tests using different materials, Kapur observed no significant difference between the grinding rate constant of classes above P_k from a fresh sample and the circulating load in the standard Bond test. He suggested that the grinding rate constant from the second grinding cycle of the standard Bond test could be used to estimate the w_i using Equation (6):

$$w_{i,K} = 1.1 \cdot 2.648 \cdot P_{100}^{0.406} \cdot k_2^{-0.81} \cdot (X \cdot M)^{-0.853} \cdot (1-X)^{-0.099}, (kWh/t) \tag{6}$$

wherein:
P_{100}—closing sieve size, (μm);
k_2—grinding rate constant of class $+P_k$ from the second grinding cycle of the standard Bond test:

$$k_2 = \frac{\ln[M - Z_1 \cdot (1-X)] - \ln(M - Z_2)}{N_2} \tag{7}$$

wherein:
X—participation of size class over P_{100} in the initial sample, (partial unit);
M—mineral charge in the mill, (g);
Z_1 and Z_2—weight of the under-size in the first and grinding cycle, (g);
N_2—number of revolutions of the mill in the second grinding cycle, (rev).

Numerical coefficients and exponents in Equation (6) were determined using the least-squares method, provided that the differences between the estimated and experimental values of w_i were minimised. The mean square relative error reported between the values of the standard method w_i and $w_{i,K}$ was 9.7%. Kapur stated that this abbreviated test does not substitute for the standard Bond test, recommending it for daily ore grinding monitoring for control purposes

Karra [14] developed a mathematical algorithm for simulating the Bond test based on the first two grinding cycles from the standard test. It can be considered a modified procedure of the one proposed by Kapur [13]. He considered that the circulating load in the standard Bond test has lower grindability and shows slower grinding behaviour. The Bond test is simulated until a circulating load of 250% is established. The value G (g/rev) is obtained from the last simulated grinding cycle, but P_{80} (μm) cannot be estimated. Therefore, in this procedure, the Bond formula cannot be used to calculate the work index, but the empirical formula obtained by statistical data processing can be used. The Karra algorithm is performed using the first two cycles of the standard Bond test and then determining the estimated value:

M—sample mass, (g);
$C = \frac{M}{3.5}$—desired under-size mass of the closing screen size at steady state, (g);
F_{80}—80% passing feed particle size, (μm);
Y—class participation $(-P_{100} + 0)$ in the starting sample, (partial unit);
Z_1 and Z_2—weight of the under-size of the closing screen size in the first and second grinding cycle, (g);
N_1 and N_2—number of mill revolutions in the first and second grinding cycle.

Further simulation is performed by calculation, provided that $M \cdot Y < C$, according to the following formulas:

$$k_1 = \frac{(1-Y)}{N_1} \cdot \left(\frac{Z_1 - M \cdot Y}{M - M \cdot Y} \right) \tag{8}$$

$$k_2 = \frac{1}{(M - Z_1) \cdot N_2} \cdot (Z_2 - Z_1 \cdot Y - Z_1 \cdot k_1 \cdot N_2) \tag{9}$$

First cycle:

$$G_1 = \frac{Z_1 - M \cdot Y}{N_1} \tag{10}$$

Second cycle:
$$G_2 = \frac{Z_2 - Z_1 \cdot Y}{N_2} \qquad (11)$$

Subsequent cycles:
$$N_i = \frac{C - Y \cdot Z_{i-1}}{G_{i-1}} \qquad (12)$$

$$Z_i = Z_{i-1} \cdot Y + Z_{i-1} \cdot N_i \cdot k_1 + (M - Z_{i-1}) \cdot N_i \cdot k_2 \qquad (13)$$

$$G_i = \frac{Z_i - Y \cdot Z_{i-1}}{N_i} \qquad (14)$$

The simulation is performed until a stable value of G (g/rev) is reached. The Bond work index is estimated by Equation (15).

$$w_{i,Kr} = 1.1 \cdot 9.934 \cdot P_c^{0.308} \cdot G^{-0.696} \cdot F_{80}^{-0.125}, \; [\text{kWh/t}] \qquad (15)$$

wherein:
P_c—closing screen size, (μm);
G—net weight of undersize product per unit revolution of the mill, (g/rev);
F_{80}—the 80% passing feed particle size, (μm).

The mean square relative error between w_i and $w_{i,Kr}$ is 4.77%, better than the Kapur algorithm.

Mular and Jergensen [15] proposed the Anaconda method, which does not require a Bond mill or a reference ore for comparison in each test. The Anaconda procedure uses a mill that is calibrated with a reference ore or ores, and the Bond work index is calculated by Equation (16):

$$w_{i,An} = \frac{A}{\left(\frac{1}{\sqrt{P_{80}}} - \frac{1}{\sqrt{F_{80}}}\right)}, \; [\text{kWh/t}] \qquad (16)$$

where in:
A—mill calibration factor, (kWh/t);
F_{80}—the 80% passing feed particle size, (μm);
P_{80}—product on milling which grindability is determined, (μm).

To determine the calibration constant A of the laboratory mill, the value of the work index w_i at a given size of the opening of the closing sieve size P_{100} should be determined on the reference ore(s) by the standard Bond test. After that, samples of the same ores should be ground in the laboratory mill at the same time t and determined for each grinding cycle F_{80} and P_{80}. Based on the obtained results, the mill calibration constant A is determined as the average value of several measurements using Equation (17).

$$A = w_i \cdot \left(\frac{1}{\sqrt{P_{80}}} - \frac{1}{\sqrt{F_{80}}}\right) \qquad (17)$$

At the Anaconda Research Center, they worked with a mill 210 mm in diameter and 251 mm long at 96% of critical speed and charged with the grinding media distribution shown in Table 1.

The feed consisted of 1 kg samples with the particle size (−1.651 + 0.147 mm). The closing sieve size was $P_{100} = 147$ μm; wet grinding was performed with 50% wt solids in the pulp for 10 min. Under these conditions, they reported $A = 0.5031$ kWh/t, so Equation (16) could be written as shown in Equation (18).

$$w_{i,An} = \frac{0.5031}{\left(\frac{1}{\sqrt{P_{80}}} - \frac{1}{\sqrt{F_{80}}}\right)}, \; [\text{kWh/t}] \qquad (18)$$

A value varies with P_{100}, the feed weight, the grinding time, and other grinding parameters. Equation (18) gives a work index estimation for $P_{100} = 147$ μm under the grinding conditions at the Anaconda Research Center. The mean square relative error between w_i and $w_{i,An}$ was reported as 4.09%., which can be considered as excellent. The procedure itself is quick and straightforward when A is known, although its determination must be performed carefully.

Table 1. Ball loading of the mill used in the Anaconda method.

Diameter of Balls, mm	Number of Balls	Mass, g
35.6–38.1	11	2316.5
31.8–33.0	17	2325.4
29.2–31.0	13	1534.8
25.4–27.9	10	822.5
24.1–25.4	7	449.7
22.9–24.1	30	1634.0
Total	88	9082.9

Nematollahi [16] proposed the estimation of w_i using a 200 mm × 200 mm mill with the grinding charge shown in Table 2.

Table 2. Characteristics of balls used by Nematollahi in the test.

Ball diameter (mm)	38.1	31.75	25.4	19.05	15.87
Number of balls	13	20	3	21	28

The initial sample volume is 207 cm^3 instead of 700 cm^3; accordingly, the test can be performed on 3 kg instead of 10 kg. The procedure involves dry grinding in a closed cycle until a 250% circulating load is reached. The Bond work index is estimated using Equation (19).

$$w_{i,N} = \frac{11.76}{p_{100}^{0.23}} \cdot \frac{1}{G^{0.75}} \cdot \frac{1}{\frac{10}{\sqrt{P_{80}}} - \frac{10}{\sqrt{F_{80}}}} \quad (19)$$

The main advantage is the lower sample mass requirement. The disadvantage of this procedure is the calibration of the mill itself. Table 3 shows the comparative values obtained between the standard Bond ball mill and the Nematollahi mill.

Menéndez-Aguado et al. [17] examined the possibility of determining the work index in a Denver laboratory batch ball mill (Figure 1) with the same inner diameter as the Bond ball standard mill. The research was performed on the size class of 100% −3.35 mm using samples of gypsum, celestite, feldspar, clinker, limestone, fluorite, and copper slag. Considering that the Bond mill/Denver mill volume ratio is 2.15, the initial sample volume was 326 cm^3 instead of 700 cm^3. Accordingly, the grinding charge was adjusted, as shown in Table 4. The grinding procedure in the Denver mill followed Bond's methodology step by step, only needing volume adjustment. The Bond work index is estimated by following Equation (20):

$$w_{i,MA} = \frac{44.5}{p_{100}^{0.23} \cdot (2.15 \cdot G)^{0.82} \cdot \left(\frac{10}{\sqrt{P_{80}}} - \frac{10}{\sqrt{F_{80}}}\right)} \quad (20)$$

Table 3. Comparative values of w_i and $w_{i,N}$.

Sample	Bond Mill w_i (kWh/t)	Nematollahi Mill $w_{i,N}$ (kWh/t)	Difference (%)
Barite	6.12	6.21	1.47
Feldspar	11.75	11.12	−5.36
Hematite	13.89	14.31	3.02
Calcite	8.36	8.50	1.67
Chromite	14.98	15.70	4.81
Dolomite	21.77	19.18	−11.90
Coke	30.43	28.75	−5.52
Coal	12.99	12.33	−5.08
Silica	11.93	11.49	−3.69
Fluorite	7.40	7.28	−1.62
Magnetite	9.33	9.54	2.25
Mean-square relative error			5.09

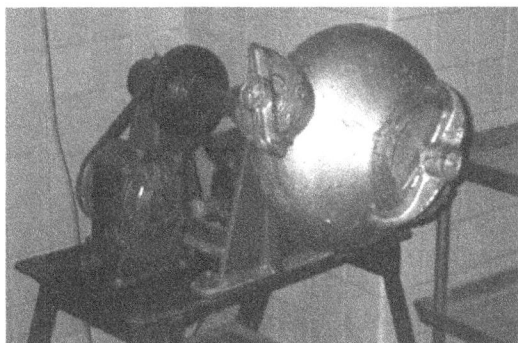

Figure 1. Denver laboratory batch ball mill.

Table 4. Ball charge in Bond and Denver mills used by Menendez Aguado et al.

Bond Mill (1952)			Denver Mill		
Number of Balls	Diameter, cm	Mass, g	Number of Balls	Diameter, cm	Mass, g
22	3.810	5951	10	3.810	2705
34	3.175	4767	16	3.175	2243
50	2.540	3750	23	2.540	1725
54	2.223	3007	25	2.223	1393
73	1.905	2920	34	1.905	1360
Total: 233	Total:	20396	Total:108	Total:	9426

The main advantages of this procedure are the availability of the Denver mill and the lower initial mass requirement. Table 5 shows the comparative values reported, showing a mean square relative error of 3.71%.

Table 5. Comparative values of w_i and $w_{i,MA}$.

Sample	P_{100} (µm)	w_i (kWh/t)	$w_{i,MA}$ (kWh/t)	Difference (%)
Limestone	200	8.99	9.05	−0.67
Feldspar	200	11.06	10.92	+1.27
Celestite	200	5.41	5.57	−2.96
Clinker	200	12.36	12.25	+0.89
Gypsum	200	6.08	5.78	+4.93
Fluorspar	200	6.94	7.41	−6.77
Copper slag	200	18.40	19.10	−3.80
Mean-square relative error				+3.71

Mucsi [18] presented a relatively fast method for estimating w_i for brittle materials (limestone, crushed pebble, bauxite, zeolite and basalt) using a Hardgrove mill with a torque meter (measuring cell), enabling the direct measurement of the power delivered to the mill. The test requires 50 g of the initial sample size of 1180–600 µm, the load of the ring on the grinding tip to be 290 N, and the grinding time to be 3 min (60 revolutions of the mill at a speed of 20 rpm). The closing screen size is 75 µm, and the Hardgrove index is determined by Equation (21):

$$H = 13 + 6.93 \cdot m_H \tag{21}$$

wherein:

m_H—weight under 75 µm;

H—Hardgrove index.

The Bond work index can be estimated from the Hardgrove index using Equation (22):

$$w_{i,H} = \frac{435}{H^{0.82}} \tag{22}$$

The Hardgrove index is based on fine products for a given number of revolutions, and the Bond work index is based on the mass of a fine product multiplied by the number of revolutions of the mill. However, it should be taken into account that other factors may also influence the given input torque in the grinding mill, such as friction, cohesion, adhesion, and material volume flow characteristics. These parameters are taken into account when specific grinding is measured by measuring torque in the manner described previously. Specific energy consumption ($W_{s,H}$) is calculated using Equation (23) when grinding is performed in a universal Hardgrove mill. The measurement of no-load energy (torque) must be subtracted from the total measured energy to determine only the specific shredding energy.

$$W_{s,H} = \frac{\int_0^\tau 2\pi n [M(t) - M_0] dt}{m} \tag{23}$$

wherein:

$M(t)$—torque (balls + material) (Nm);

M_0—no-load torque (Nm);

n—revolution per min (1/s);

t—grinding time (s);

m—mass of sample below 75 µm (g).

The specific energy consumption for comminution in the Bond mill is calculated by Equation (24);

$$W_{s,B} = \frac{W_{balls+material} - W_0}{m_p} \tag{24}$$

wherein:

$W_{balls+material}$—measured work with ball and mineral charge (kWh);
W_0—measured work without charge (kWh);
m_p—the product mass (t).

Once $W_{s,B}$ or $W_{s,H}$, P_{80}, and F_{80} are known, w_i can be estimated after isolating it in Equation (1) for both cases.

The reported relative difference values between w_i and $w_{i,H}$ for different ores ranged from −8.1 to 24.1%. The advantages of this method are the use of a simple well-controlled laboratory mill, the need of only 50 g of sample, and a short testing time (60–90 min).

Saeidi et al. [19] rely on the mill designed by Nematollahi [16] to determine the Bond work index, estimated by Equation (19). Using a representative sample (iron ore) of 2 kg, the PSD was determined, and the representative sample was then ground at time intervals of 20, 60, 120, and 180 s. After each grinding, the sample was sieved, P_{80} was obtained, and the work index was determined. The obtained results are shown in Table 6.

Table 6. Comparative results for w_i versus $w_{i,N}$.

Grinding Time (s)	w_i (kWh/t)	$w_{i,N}$ (kWh/t)	Difference (%)
20	13.1	8.21	37.33
60	12.36	8.04	34.95
120	11.68	7.84	32.88
180	11.11	7.81	29.70

Due to large deviations shown in Table 6, the authors defined a new Equation (27) to determine the Bond work index based on the obtained results. They came up with a new formula by examining the relationship between the parameters G (g/rev) and P_{80} (μm) and the grinding time for this ore, resulting in Equations (25) and (26):

$$P_{80} = -0.1085 \cdot t + 122.56 \tag{25}$$

$$G = -1E - 0.6 \cdot t^2 + 0.0004 \cdot t + 0.3397 \tag{26}$$

Finally, w_i can be estimated using Equation (27), which is the result of combining Equation (25) and Equation (26) with Equation (19), resulting in a new equation to estimate the Bond work index:

$$w_{i,SA} = \frac{5.6}{(-1E - 0.6 \cdot t^2 + 0.0004 \cdot t + 0.3397)^{0.75}} \cdot \frac{1}{\frac{10}{\sqrt{-0.1085 \cdot t + 122.56}} - \frac{10}{\sqrt{F_{80}}}} \tag{27}$$

In order to determine the accuracy of Equation (27), an additional grinding run of 100 s was performed, and the results of which are shown in Table 7.

Table 7. Comparison of results for w_i versus $w_{i,SA}$.

Grinding Time (s)	w_i (kWh/t)	$w_{i,SA}$ (kWh/t)	Difference (%)
100	12.18	12.13	0.41

Mwanga et al. [20] developed a Geometallurgical Comminution Test (GCT) that requires a small amount of initial sample and a jar mill (Capco, Ipswich, UK). The grinding test is performed on a sample under 3.35 mm with a starting weight of 220 g and can be performed within 2–3 h. The sample is ground while dry for 2, 5, 10, 17, and 25 min. After each grinding time, the PSD is determined by sieving, and the sample is returned to the mill for further grinding. P_{80} is obtained from the PSD, and the power consumption is measured during the test. When the test is performed at a constant sample mass and mill parameters (number of revolutions, grinding batch), it can be assumed that the energy supplied to the mill per unit time is constant. From Equation (1), for the given feed size, the change in specific grinding energy is proportional to the reciprocal of the square root of P_{80}:

$$W \cdot \frac{\sqrt{P_{80}}}{10} = \text{constant} \Rightarrow w_{i,GCT} = W \cdot \frac{\sqrt{P_{80}}}{100} \qquad (28)$$

Comparing the results from the two grindability tests revealed that there is a linear relationship between the work indices. The model for estimating the Bond work index from the GCT test data is then given by Equation (29):

$$w_i = w_{i,GTC} \cdot \left(\frac{1}{\sqrt{\lambda}} \cdot \eta \cdot 1/4\right) \qquad (29)$$

wherein:

λ—geometric division factor and is $\lambda = 2.65$;
η—mill drive and engine efficiency and amounts to $\eta = 0.64$;
$W_{i,GTC}$—operating index of the GCT, calculated using Equation (28).

The authors stated that the test and performance of the presented method were confirmed on several ores, with the relative error ranging from 0.70% to f 8.8%. The advantages of this method are the small sample quantity that is needed (220 g of the initial sample) and the short testing time (results can be obtained in 25 min, taking the entire test no more than three hours). The disadvantage is the availability of the mill itself; the authors recognised that the proposed method does not aim to substitute the standard Bond test.

Lewis et al. [21] developed a new method of grinding testing based on computer simulation, closely related to the standard Bond method. The simulation is based on a mathematical algorithm that simulates a standard Bond test and is divided into two parts. The first part uses experimental data from the first grinding cycle to obtain the initial parameters of the model. The calculated parameters and raw material characteristics are stored in a database to be used in the second part of the simulation for prediction purposes. The prediction method simulates a standard test. For each grinding cycle, all raw material that is smaller than the opening of the comparative sieve is replaced by a representative mass of the starting sample. The calculation continues using the parameter values set for a given grinding cycle. Four grinding cycles are calculated automatically. A check is performed during the fourth and any subsequent grinding cycles to assess whether the newly formed undersize mass per mill revolution G (g/rev) is constant (within 3%) for the last three grinding cycles. If G (g/rev) is constant, a steady state is reached; otherwise, the computer procedure continues with the next grinding cycle. When a steady state is reached, the Bond work index is calculated using Equation (32). The mean square relative difference between the values of the Bond work index obtained by the standard method and the values obtained by computer simulation is 2.81%.

Aksani and Sönmez [22] proposed a computer simulation of the Bond grind test using a cumulative kinetic model [23,24]. The model contains only two parameters, which simplifies the interpretation of the results. Equation (30) gives the relationship between the comminution speed and the particle size:

$$k = C \cdot x^n \qquad (30)$$

wherein:

k—breakage rate constant (min^{-1});
C and n—constants that are dependent on the mill and material characteristics;
x—sieve size (μm).

A standard Bond mill and a standard Bond grind test were used to determine the model parameters. The test is performed on a sample of mass M (g) that is 700 cm^3 of size class -3.35 mm. The sample is ground at times of 0.5 min, 1 min, 2 min, and 4 min. After each grinding cycle, the analysis of the PSD is determined for the sample.

The PSD products are combined and returned to the mill for the subsequent grinding cycle. The grinding rate constant k is calculated by nonlinear regression using the obtained cumulative reflection data in relation to the grinding time. To calculate the parameters C and n, Equation (30) should be logarithmic, and then linear regression should be applied. The computer simulation uses PSD data, initial input mass, kinetic model parameters, and mill speed for the first grinding cycle. The prediction test simulates the standard Bond procedure. After each grinding cycle, the newly formed undersize mass per mill revolution G (g/rev) is calculated, and the material under P_{100} is replaced with the same mass of feed sample. The calculation continues until G (g/rev) becomes constant for the last three grinding cycles. When a steady state is reached, the parameters obtained in the last grinding cycle and Equation (32) are used to calculate the Bond work index. The mean square relative error between the values of the Bond work index obtained by the standard method and the values obtained by computer simulation was 2.54%.

Ford and Sithole [25] provided an abbreviated method for w_i estimation consisting of two tests. The first test was performed with only one grinding cycle, and the second test was performed with three grinding cycles.

In the first test, a sample of 700 cm^3 with a size of 100% under 3.35 mm is ground in a standard Bond ball mill in time intervals of 0.5 min, 1 min, 2 min, and 4 min. After each grinding run, the mass of the sample is measured, and the PSD is determined. These data are then used to calculate the parameter k for each size x (see Equation (31)).

$$W_{(x,t)} = W_{(x,0)} \cdot \exp(-k \cdot t) \tag{31}$$

wherein:

t—grinding time (min);
$W_{(x,t)}$—cumulative content of screening aperture x during grinding t;
$W_{(x,0)}$—cumulative reflection content of the initial sample for the sieve opening x;
k—breakage rate constant (min^{-1}).

The model describes a mathematical simulation in a closed grinding cycle. In the simulation, the number of revolutions varies until a circulating load of 250% is reached. The parameters G (g/rev), P_{80} (μm), and F_{80} (μm) are estimated using simulation, and the Bond work index is estimated using the standard method Bond equation (Equation (32)) using the simulated parameters (Equation (33)).

$$w_i = \frac{44.5}{p_{100}^{0.23} \cdot (G)^{0.82} \cdot \left(\frac{10}{\sqrt{P_{80}}} - \frac{10}{\sqrt{F_{80}}}\right)} \tag{32}$$

$$w_{i,FS1} = \frac{44.5}{p_{100}^{0.23} \cdot (G_s)^{0.82} \cdot \left(\frac{10}{\sqrt{P_{80,s}}} - \frac{10}{\sqrt{F_{80,s}}}\right)} \tag{33}$$

wherein G_s, $F_{80,s}$, and $P_{80,s}$ are G (g/rev), P_{80} (μm), and F_{80} (μm) obtained by simulation, respectively.

The feature of this method is that the work indices can be simulated for different PSD based on the results of only one grinding cycle.

The second proposed test is based on the standard Bond test, considering only the first three grinding cycles. After the third cycle, G (g/rev) and P_{80} are determined and used to calculate the Bond work index via Equation (34):

$$w_{i,FS2} = \frac{44.5}{p_{100}^{0.23} \cdot (G_3)^{0.82} \cdot \left(\frac{10}{\sqrt{P_{80,3}}} - \frac{10}{\sqrt{F_{80,3}}}\right)} \tag{34}$$

wherein G_3, $F_{80,3}$, and $P_{80,3}$ are G (g/rev), P_{80} (μm), and F_{80} (μm) obtained experimentally after only three cycles of the Bond standard test.

The mean square relative error between w_i and $w_{i,FS1}$, and w_i and $w_{i,FS2}$ resulted in 11.71%, and 2.20%, respectively. The second procedure takes longer but leads to better results than the first one.

Gharehgheshlagh [26] presented a method for calculating the Bond work index that tracks the grinding kinetics in a Bond ball mill. The method is fast and practical because it establishes a relationship between the grinding parameters and the parameters of the Bond equation and eliminates specific steps of the laboratory test due to the reduction of the grinding cycle. The test is performed by grinding 700 cm³ of a sample 100% under 3.35 mm in a Bond ball mill for 0.33, 1, 2, 4, and 8 min. After each grinding cycle, the grinding product PSD is determined and returned to the mill for the subsequent grinding cycle. This grinding kinetics analysis is used to determine the functional dependence between the number of mill revolutions and undersize mass passing P_{100} (m$_{us}$) as well as the relationship between the number of mill revolutions and P_{80} (μm) using the least-squares numerical method. The first function determines the number of mill revolutions $N_{250\%}$ (revolutions) required to obtain the under-size mass, such that the circulating load is 250%. Based on the values of $N_{250\%}$ (revolutions) and the determined functional dependencies, the parameters G (g/rev) and P_{80} (μm) are estimated, and Equation (32) can be used to estimate the work index. The mean square relative error between the real and estimated w_i was 1.23%.

Ciribeni et al. [27] introduced a Bond test simulation based on the cumulative kinetic model [23,24]. The simplified procedure consists of calculating the kinetic parameters after only one grinding run, instead of a series of runs. Finally, the estimation of w_i is performed through mathematical simulation. The test is performed by grinding a 700 cm³ sample, 100% under 3.35 mm, in a Bond ball mill for 5 min. The kinetic parameters are determined: k by Equation (30) and C and n by Equation (35).

$$\ln(k) = \ln(C) + n \cdot \ln(x) \tag{35}$$

Once estimated by the simulation of G and P_{80}, the Bond work index is estimated using Equation (32). Several ores were used for validation, and the mean square relative error between standard and calculated work index was reported as 6.31%.

Magdalinović [28] presented an abbreviated test for determining the work index based on performing two grinding cycles and relying on the law of first-order kinetics. The test is performed on a sample prepared 100% under 3.35 mm, in a standard Bond mill. Feed sample PSD is obtained, the initial sample mass of 700 cm³, M (g), is determined, and the grinding product mass at steady-state, IGP, is calculated with Equation (36):

$$IGP = M/3.5 \ [g] \tag{36}$$

The first grinding cycle feed is prepared with the IGP weight of the initial sample and made equal to M weight with the initial sample after removing by sieving the undersize of P_{100}. This composite sample is ground for an arbitrary number of mill revolutions (N_1, usually 50, 100, or 150), and the oversize mass and the undersize mass are weighed. The oversize grinding rate constant (k) can be calculated using Equation (37);

$$k = n \cdot \frac{\ln R_0 - \ln R_1}{N_1} \tag{37}$$

wherein:

R_0—oversize in the initial sample (%);
R_1—oversize in the product of cycle 1 (%);
n—number of revolutions per min, (min^{-1});
N_1—total number of mill revolutions in cycle 1.

Once the grinding rate constant k is determined, it is used to obtain the necessary mill revolutions N_2 required to obtain a circulating load of 250%. The second cycle feed is obtained as it was in the first cycle. The second cycle involves a grinding operation for

N_2 revolutions. Once the cycle is finished, the product is sieved at P_{100}. The undersize mass should be approximately IGP. The G value (g/rev) is calculated, and P_{80} is obtained from the PSD. Equation (32) can be used to estimate the Bond work index. The mean square relative error between the actual and the calculated values of the work index by the Magdalinović test with two grinding cycles was 4.9%. In 2003, Magdalinović [29] proposed the abbreviated test with three grinding cycles by adding one cycle to the procedure proposed in 1989. In this case, the mean square relative error diminished to 2.75%. As expected, a lower minor error is obtained with the abbreviated three cycle test.

Todorovic et al. [30] proposed an abbreviated method that could be done with two, three, or four grinding cycles. Each grinding cycle is done in the same way as in the standard Bond procedure. In the shortened procedure with two grinding cycles, the PSD of the initial sample, F_{80} (μm), and X (the oversize mass at P_{100}) are determined. From the feed, which must be prepared 100% under 3.327 mm, a sample of 700 cm³ is taken, and its weight M (g) is determined. This sample is ground in a Bond ball mill for an arbitrary number of revolutions (N_1 = 50, 100, or 150 revolutions). Afterwards, the grinding product is sieved at P_{100} and R (retained weight, g) and D (undersize weight, g) are determined. The undersize weight D is the sum of the undersize in the fresh feed D_u (g) and the newly formed undersize D_n. The newly formed undersize mass $D_n = D - D_u$ is calculated. In the first cycle, $D_u = M \cdot (1-X)$ (g), while in the subsequent cycle, $D_u = D_{n-1} \cdot (1-X)$ (g), wherein D_{n-1} is the undersize mass of the sieves from the previous cycle, (g). The newly formed undersize mass per mill revolution G (g/rev) is then calculated, and the number of mill revolutions for the subsequent grinding cycle N_n is according to Equation (38).

$$N_n = \frac{\frac{M}{3.5} - D_{(n-1)} \cdot (1-X)}{G} \quad [\text{rev}] \tag{38}$$

A fresh feed sample equaling D_{n-1} is blended with the retained material from the previous cycle, fed into the mill, and ground for N_n revolutions. The grinding product is again sieved, and the retained product is weighed to obtain the R of this cycle. The constant k is then calculated using Equation (39):

$$k = \frac{n \cdot (\ln R_0 - \ln R)}{N} = \frac{n \cdot \left[\ln\left(\frac{R_{(n-1)}}{M} + \frac{D_{(n-1)}}{M} \cdot X\right) - \ln \frac{R}{M}\right]}{N} \tag{39}$$

The required number of revolutions N is calculated to produce the steady-state weight undersize at 250% circulating load (Equation (40)):

$$N = \frac{n}{k}\left[\ln\left(\frac{2.5}{3.5} \cdot 100 + \frac{X}{3.5} \cdot 100\right) - \ln\left(\frac{2.5}{3.5} \cdot 100\right)\right] \tag{40}$$

The parameter G (g/rev) is calculated using Equation (41), and the last cycle G_e value is estimated using Equation (42). The P_{80} value in this cycle is estimated using Equation (43).

$$G = \frac{\frac{M}{3.5} \cdot X}{N} \tag{41}$$

$$G_e = 1.158 \cdot G \quad (\text{g/rev}) \tag{42}$$

$$P_{80} = 1.035 \cdot P_{80,n-1} \quad (\mu m) \tag{43}$$

Using the values of G_e (g/rev) and P_{80} (μm) in Equation (32), the estimation of the work index $w_{i,T}$ (kWh/t) is obtained. Table 8 shows the results reported by Todorovic et al. [30] on mixtures of limestone and andesite, comparing w_i with $w_{i,T}$ obtained by the abbreviated test with two, three, and four grinding cycles. The mean square relative error ranged from 3.0% to 5.2%.

Table 8. Comparative values of the Bond work index according to the standard Bond test and the abbreviated test by Todorovic et al.

Sample	P_{100} (µm)	w_i (kWh/t)	$w_{i,T}$ (kWh/t)			Difference (%)		
			II Cycle	III Cycle	IV Cycle	II Cycle	III Cycle	IV Cycle
Limestone:Andesite 0:100	74	18.09	17.53	17.37	17.90	−3.09	−3.98	−1.05
Limestone:Andesite 25:75	74	17.03	17.69	16.73	16.80	3.87	−1.75	−1.33
Limestone:Andesite 50:50	74	15.15	15.58	14.89	15.03	0.51	−3.93	−3.02
Limestone:Andesite 75:25	74	14.51	14.39	13.86	14.03	−0.82	−4.48	−3.34
Limestone:Andesite 100:0	74	13.90	14.50	15.14	14.53	4.32	8.93	4.52
		Mean relative square error				2.97	5.18	2.92

A summary of the mean relative square error reported by the authors in each proposal is depicted in Figure 2. Considering the relative error values and the simplification of the laboratory procedure, the method proposed by Horst and Bassarear and the one by Gharehgheshlagh are advantageous.

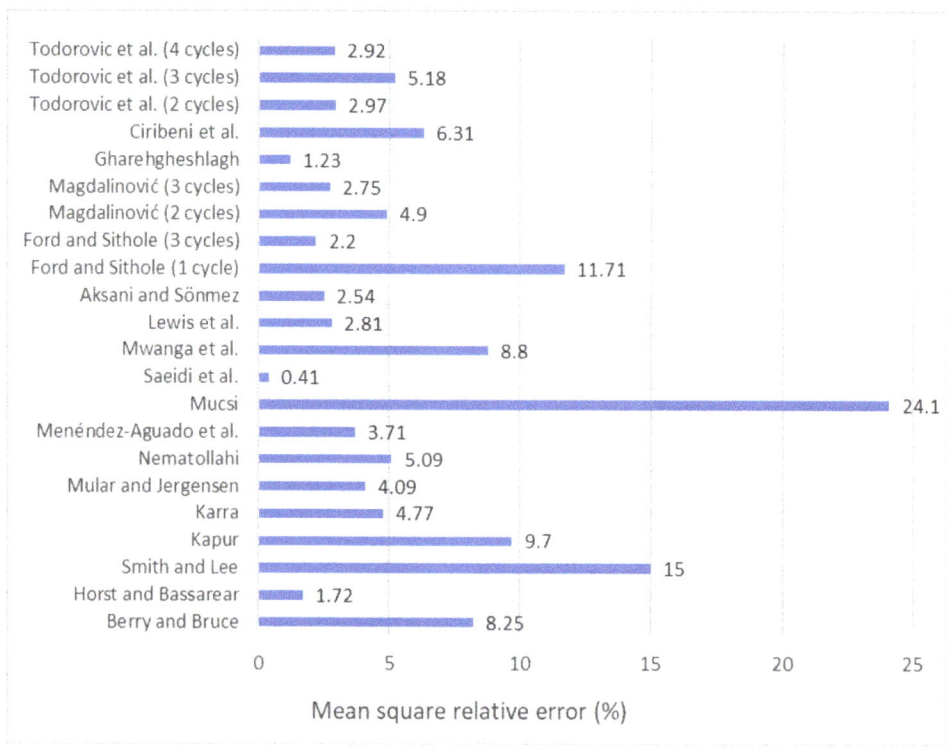

Figure 2. Summary of relative errors of alternative procedures.

3. Conclusions

Alternative abbreviated and simplified procedures for determining the work index have been proposed through the years. This review presented alternative shorter, simplified, and faster procedures that can be classified into two groups:

1. Alternative tests that simulate the standard Bond test with an abbreviated procedure;
2. Alternative tests based on determining problem sample grindability using a reference sample with w_i known.

Alternative tests from the first group are based on the use of a Bond standard ball mill for the reach the steady-state more quickly [28–30] or for performing the mathematical simulation of the standard test [13,14,21,22,27].

Alternative tests from the second group can be performed in a different mill, usually needing less sample than the standard procedure. All of the methods aim to give a close estimation of the Bond work index when the standard Bond ball mill is not available and are faster procedures with a reduced number of grinding steps. The longest alternative test requires 3–4 grinding cycles, while the shortest one can be performed with one grinding cycle. It must be considered that the standard procedure compels a minimum of 5 grinding cycles, with 7–10 grinding cycles usually being necessary.

In general, the mean square error data presented cannot be understood as a validity indicator, for in some cases, the reported value was based on just a few tests or with few ores. However, these data indicate that shorter procedures (i.e., with just one grinding cycle) are usually less reliable, yielding a higher mean square error. Nevertheless, due to the advantage in laboratory time, they could be recommended if ore feed is the same, which could be the case of the periodic grindability control in a specific mine.

Finally, after an adequate grinding kinetic behaviour characterisation of the ore, alternative tests based on the simulation of the standard Bond test could be recommended when considering the process digitalisation as part of the global digitalisation strategy in the mining industry.

Author Contributions: Conceptualisation, V.N. and J.M.M.-A.; methodology, G.G.G. and J.M.M.-A.; investigation, V.N., M.S.T., G.G.G. and A.L.C.-V.; resources, V.N., M.T., M.S.T. and J.M.M.-A.; writing—original draft preparation, V.N., M.T. and M.S.T.; writing—review and editing, A.L.C.-V., G.G.G. and J.M.M.-A.; visualisation, V.N., M.S.T., G.G.G. and J.M.M.-A.; supervision, M.T. and A.L.C.-V. All authors have read and agreed to the published version of the manuscript.

Funding: This research did not receive external funding.

Institutional Review Board Statement: Not applicable.

Informed Consent Statement: Not applicable.

Data Availability Statement: Not applicable.

Conflicts of Interest: The authors declare no conflict of interest.

References

1. Bond, F.C. Third theory of comminution. *Trans. AIME Min. Eng.* **1952**, *193*, 484–494.
2. GMG-Global Mining Guidelines Group. Determining the Bond Efficiency of Industrial Grinding Circuits. 2016. Available online: https://gmggroup.org/wp-content/uploads/2016/02/Guidelines_Bond-Efficiency-REV-2018.pdf. (accessed on 30 May 2021).
3. García, G.G.; Oliva, J.; Guasch, E.; Anticoi, H.; Coello-Velázquez, A.L.; Menéndez-Aguado, J.M. Variability Study of Bond Work Index and Grindability Index on Various Critical Metal Ores. *Metals* **2021**, *11*, 970. [CrossRef]
4. Nikolić, V.; Trumić, M. A new approach to the calculation of bond work index for finer samples. *Miner. Eng.* **2021**, *165*, 106858. [CrossRef]
5. Narayanan, S.S.; Whiten, W.J. Determination of comminution characteristics from single particle breakage tests and its application to ball mill scale-up. *Trans. Inst. Min. Metall.* **1988**, *97*, C115–C124.
6. Powell, M.; Hilden, M.; Ballantyne, G.; Liu, L.; Tavares, L.M. The Appropriate, and Inappropriate, Application of the JKMRC t10 Relationship. In Proceedings of the XXVII International Mineral Processing Congress IMPC, Santiago, Chile, 20–24 October 2014.
7. Ballantyne, G.; Peukert, W.; Powell, M.S. Size specific energy (SSE)—Energy required to generate minus 75 micron material. *Int. J. Miner. Process.* **2015**, *136*, 2–6. [CrossRef]

8. Lvov, V.V.; Chitalov, L.S. Comparison of the Different Ways of the Ball Bond Work Index Determining. *Int. J. Mech. Eng. Technol.* **2019**, *10*, 1180–1194.
9. Berry, T.F.; Bruce, R.W. A simple method of determining the grindability of ores. *Can. Min. J.* **1966**, *87*, 63–65.
10. Coello Velázquez, A.L.; Menéndez-Aguado, J.M.; Brown, R.L. Grindability of lateritic nickel ores in Cuba. *Powder Technol.* **2008**, *182*, 113–115. [CrossRef]
11. Horst, W.E.; Bassarear, J.H. Use of simplified ore grindability technique to evaluate plant performance. *Trans. Metall. Soc. AIME* **1977**, *260*, 348–351.
12. Smith, R.W.; Lee, K.H. A comparison of data from Bond type simulated closed-circuit and batch type grindability tests. *Trans. Metall. Soc. AIME* **1968**, *241*, 91–99.
13. Kapur, P.C. Analysis of the Bond grindability test. *Trans. Inst. Min. Metall.* **1970**, *79*, 103–107.
14. Karra, V.K. Simulation of Bond grindability tests. *CIM Bull.* **1981**, *74*, 195–199.
15. Mular, A.L.; Jergensen, G.V. *Design and Installation of Comminution Circuits, Society of Mining Engineers of the American Institute of Mining*; Metallurgical, and Pegroleum Engineers, Inc.: New York, NY, USA, 1982; p. 1022.
16. Nematollahi, H. New size laboratory ball mill for Bond work index determination. *Min. Eng.* **1994**, *46*, 352–353.
17. Menéndez-Aguado, J.M.; Dzioba, B.R.; Coello-Velazquez, A.L. Determination of work index in a common laboratory mill. *Miner. Metall. Process.* **2005**, *22*, 173–176. [CrossRef]
18. Mucsi, G. Fast test method for the determination of the grindability of fine materials. *Chem. Eng. Res. Des.* **2008**, *86*, 395–400. [CrossRef]
19. Saeidi, N.; Noaparast, M.; Azizi, D.; Aslani, S.; Ramadi, A. A developed approach based on grinding time to determine ore comminution properties. *J. Min. Environ.* **2013**, *4*, 105–112.
20. Mwanga, A.; Rosenkranz, J.; Lamberg, P. Development and experimental validation of the Geometallurgical Comminution Test (GCT). *Miner. Eng.* **2017**, *108*, 109–114. [CrossRef]
21. Lewis, K.A.; Pearl, M.; Tucker, P. Computer Simulation of the Bond Grindability test. *Miner. Eng.* **1990**, *3*, 199–206. [CrossRef]
22. Aksani, B.; Sönmez, B. Simulation of bond grindability test by using cumulative based kinetic model. *Miner. Eng.* **2000**, *13*, 673–677. [CrossRef]
23. Finch, J.A.; Ramirez-Castro, J. Modelling Mineral Size Reduction in Closed-Circuit Bal Mill at the Pine Point Mines Concentrator. *Int. J. Miner. Process.* **1998**, *8*, 67–78.
24. Ersayin, S.; Sönmez, B.; Ergün, L.; Aksarri, B.; Erkal, F. Simulation of the Grinding Circuit at Gümtüşköy Silver Plant. *Turk. Trans. IMM* **1993**, *102*, C32–C38.
25. Ford, E.; Sithole, V. A Comparison of Test Procedures for Estimating the Bond Ball Work Index on Zambian/DRC Copper-Cobalt Ores and Evaluation of Suitability for Use in Geometallurgical Studies, Copper Cobalt Africa. In Proceedings of the 8th Southern African Base Metals Conference, Livingstone, Zambia, 6–8 July 2015; pp. 65–68.
26. Gharehgheshlagh, H.H. Kinetic grinding test approach to estimate the ball mill work index. *Physicochem. Probl. Miner. Process.* **2015**, *52*, 342–352.
27. Ciribeni, V.; Bertero, R.; Tello, A.; Puerta, M.; Avellá, E.; Paez, M. Menéndez-Aguado, J.M. Application of the Cumulative Kinetic Model in the Comminution of Critical Metal Ores. *Metals* **2020**, *10*, 925. [CrossRef]
28. Magdalinović, N. A procedure for rapid determination of the Bond work index. *Int. J. Miner. Process.* **1989**, *27*, 125–132. [CrossRef]
29. Magdalinović, N. Abbreviated test for quick determination of Bond's Work index. *J. Min. Metall.* **2003**, *39*, 1–10.
30. Todorovic, D.; Trumic, M.; Andric, L.; Milosevic, V.; Trumic, M. A quick method for bond work index approximate value determination. *Physicochem. Probl. Miner. Process.* **2017**, *53*, 321–332.

Article

Application of the Cumulative Kinetic Model in the Comminution of Critical Metal Ores

Victor Ciribeni [1], Regina Bertero [1], Andrea Tello [1], Matías Puerta [1], Enzo Avellá [1], Matías Paez [1] and Juan María Menéndez Aguado [2,*]

[1] Instituto de Investigaciones Mineras, Universidad Nacional de San Juan, 5400 San Juan, Argentina; ciribeni@unsj.edu.ar (V.C.); reginabertero@gmail.com (R.B.); act8383@gmail.com (A.T.); matias_sp_79@hotmail.com (M.P.); enavella.91@gmail.com (E.A.); matias.p043@gmail.com (M.P.)
[2] Escuela Politécnica de Mieres, Universidad de Oviedo, 33600 Oviedo, Spain
* Correspondence: maguado@uniovi.es; Tel.: +34-985458033

Received: 10 June 2020; Accepted: 8 July 2020; Published: 9 July 2020

Abstract: Over the last decades, several reliable mathematical models have been developed for simulating ore comminution processes and determining the Work Index. Since Fred Chester Bond developed the Work Index standard procedure in 1961, numerous attempts have been made to find simpler, faster, and economically more advantageous alternative tests. In this paper, a Bond test simulation based on the cumulative kinetic model (CKM) has been checked on a spreadsheet. The research has been accomplished by conventionally determining the kinetic parameters for some Ag and Au ores and for three pure minerals and one rock that are common constituents of the gangue rock. Analysis of the results obtained allowed to develop a simplified procedure for calculating the kinetic parameters and their application to Work Index determination through simulation.

Keywords: comminution; simulation; work index; grindability; critical metals

1. Introduction

Energy consumption during the comminution stage has a severe impact on the operational costs of ore processing plants, being a key factor in operation planning and optimization. This situation has drawn the interest of researchers on early stages of ore processing, who tie the amount of energy consumed with the work done in the comminution of the mineral species involved.

The First Law of Comminution or Rittinger's Law [1] dates back to 1867 and postulates that the energy required in the mineral breakage is directly proportional to the new surface area produced. Later, in 1885, the Second Law of Comminution was postulated by Kick [2], who stated that the energy supplied is proportional to the particle volume, regardless of the original size. In 1952, Fred Chester Bond [3,4] postulated the Third Law of Comminution (also known as Bond's Law). It states that the energy required is proportional to the length of crack initiating breakage. However, as mentioned by Jankovic et al. [5], the application of Kick's and Rittinger's laws has been met with varied success and are not realistic for designing size reduction circuits, while Bond's Third Law can be reasonably applied to the range in which ball/rod mills operate. In spite of its empirical basis, Bond's Law is the most widely used method for the sizing of ball/rod mills and has become a standard. Nevertheless, despite being unrivalled, it has an error range of up to 20%. Besides, determining the Work Index (wi) for a given mineral or ore composition is a time-consuming procedure that requires qualified personnel and a significant quantity of sample [6–8].

These drawbacks were resolved to some extent by several researchers, who developed alternative methodologies to determine energy consumption during crushing and grinding [9–15]. Some methodologies [7,16–20] employed mathematical simulations based on tested mathematical models. In all cases, they involve an adequate characterization of the material in the laboratory, followed by the simulation of the grinding and sorting operations of the standard test to determine the Work Index (w_i).

The cumulative kinetic model (CKM) was developed by Laplante et al. [21], and it represents a simple solution to the basic equation proposed by Loveday [22]. As Menéndez-Aguado [8] pointed out, it possesses several advantages, the main of which are that the model is defined by two parameters, simplifying the interpretation of the results and that the parameters determined in the laboratory can be applied at industrial scale [23].

2. Theoretical Background

2.1. The Cumulative Based Kinetic Model

The CKM is based in a first-order kinetic equation, in which the particle breakage rate for a given particle size interval is proportional to the mass existing in that interval. It has the particularity of being defined in terms of only two parameters, which may be determined in laboratory batch tests and then directly applied to the model.

The kinetic parameter (k) is defined as the oversize disappearing rate for a given size class, either at continuous or discontinuous grinding under a piston flow regime, and can be described as shown in Equation (1).

$$W_{(x,t)} = W_{(x,0)} \exp(-k\,t) \tag{1}$$

where

$W(x,t)$ = cumulative percent of oversize for size class x in time t.
$W(x,0)$ = cumulative percent of oversize for size class x at the feed.
k = breakage rate constant (min^{-1})
t = time (min).

The relationship between breakage velocity and particle size is shown in Equation (2):

$$k = C\,x^n \tag{2}$$

In Equation (2), C and n are constants depending on the features of the mineral and the mill, as described by Ersayin et al. [24]. They are CKM model parameters and can be determined experimentally. Provided that size distribution in the feed stream is known, C and n allow to calculate the size distribution of the product through Equation (3).

$$W_{(x,t)} = W_{(x,0)} (\exp(-C\,x^n\,t)) \tag{3}$$

2.2. Determination of the CKM Parameters C and n

Parameters C and n may be calculated simply and quickly from a small amount of sample in a laboratory mill [8]. In our case, since the objective was simulating the Bond tests to obtain w_i, the standard mill designed by Bond was used for characterization purposes.

Keeping the same feed quantity as in Bond test (700 cm^3), successive grinding cycles were done at predefined time intervals. After each cycle, a representative sample of the mineral load was extracted, and its particle size distribution (PSD) was obtained. The part of the sample above the reference size was then conveyed back to the mill and new feed was added up to the initial load before resuming the test.

The k values are calculated for several monosizes through linear regression of the retained mass accumulated for each grinding time, using the Equation (1) linearized:

$$\ln(W_{(x,t)}) - \ln(W_{(x,0)}) = k\,t \tag{4}$$

Linearizing Equation (2) and performing a new linear regression for each monosize, C and n are obtained:

$$\ln(k) = \ln(C) + n \ln(x) \tag{5}$$

Taking into account that Equation (5) is the equation of a straight line, ln(C) is the intercept on the y-axis and n is the slope. Thus, the parameters are set for applying Equation (3) to obtain the size distribution as a function of grinding time.

2.3. Simplified Procedure to Grinding Kinetic Parameters Determination

Figure 1 shows the outcome of Equation (4) as a function of grinding times for monosizes of 840, 420, 149, and 53 μm of a pure quartz sample. It can be seen that the slope (k) remains rather constant irrespective of the time considered, from intermediate values to the end (5 min) of the test. A reduction of the test duration would be important to obtain the kinetic parameters through this procedure.

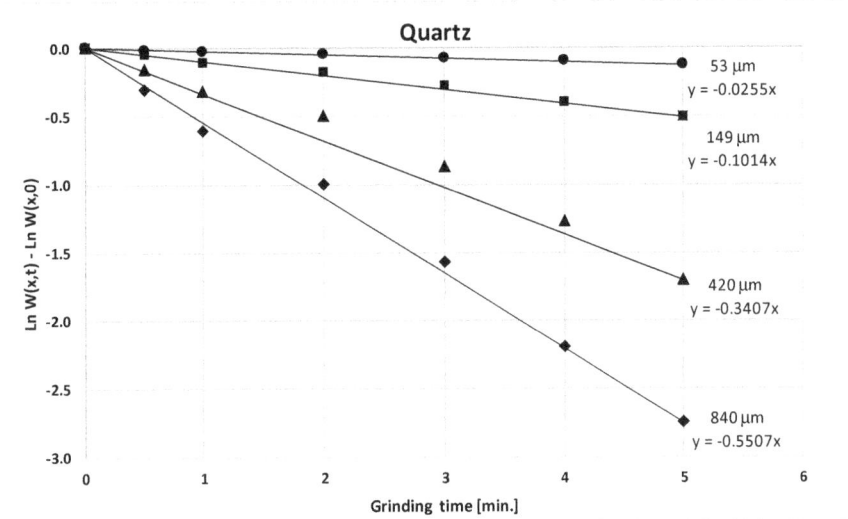

Figure 1. Determination of the kinetic constant, k, for a quartz sample; the lines are drawn from the starting point to the 5 min measurement.

To check this procedure, this work was aimed at validating the proposed methodology to determine the grinding kinetic parameters. The modelling was then used to determine the Work Index through simulation, and finally, those results were compared for eight different samples with the w_i obtained through the Bond standard test.

3. Methodology

3.1. Sample Preparation

Eight samples were selected, of which four, namely, M1S1, M2S1, M2S2, and M2S3, were critical metal ores (Ag and Au) and the rest consisted of minerals (feldspar, quartz, and calcite) and a rock (pure limestone), being usual constituents of gangue rock.

It must be taken into account that critical metal ores have commonly a very low grade, and thus, it is the gangue composition that defines their grindability behavior.

Heretofore, M1S1 stands for a hydrothermal low sulfidation mineral ore, consisting of veins and veinlets of silica (quartz, chalcedony, and opal) containing free gold, electrum, and Ag sulfosalts, in addition to cassiterite, galena, pyrite, and chalcopyrite. M2S1, M2S2, and M2S3 stand for an interpreted medium-sulfidation epithermal system, containing quartz, carbonates, and, to a lesser extent, Ag-, Au-, Pb-, Cu-, and Zn-bearing sulfides and sulfosalts.

Sample preparation was done following the usual procedures for the standard Bond test [4]: progressive comminution through several steps in jaw crusher, cone crusher, and roll/roller mills, until a final product finer than 6 mesh (3.35 mm) is obtained, avoiding an over-representation of the finest fraction.

The samples finer than 6 mesh were then subsampled with a rotary splitter. The resulting subsamples, 1 kg each approximately, were used in mineralogical, chemical, and grain size characterization and also constituted the feed and the fractions for the Bond and grinding kinetic tests.

Grain size analyses were performed after sieving, approximately 300 g of each sample in a RO-TAP Sieve Shaker using a series of ASTM sieves.

3.2. Work Index Determination

The Bond test for ball mills was performed for each sample following the abovementioned Bond standard procedure [4]. The resulting value was later used as a reference to compare with the Work Index obtained through simulation. An initial feeding sample with a volume of 700 cm^3 was weighed and then fed into the standard mill filled up with the ball load defined for the test. After a grinding period of an arbitrary number of revolutions (e.g., 100), the mill was carefully dumped, recovering the maximum possible of the fines from the ball charge and the mill liners to minimize sample losses. The material was sieved to the reference size (P_{100}), and the undersize was weighed. An equal mass of new raw material was then added to the oversize feed to compensate loss of the fines.

The process was repeated, weighing the newly produced undersize (G) concerning the reference mesh. This undersize (G) was divided by the number of revolutions resulting in the grams per revolution (Gbp). Once Gbp is known, a new grinding cycle was performed after calculation of the needed revolutions to reach the steady state. The cycle was repeated until the undersize produced per revolution (Gbp) came to an equilibrium and the circulating feed approached 250%. The Gbp of the two last cycles was then averaged to obtain the grindability index of the test. The P_{80} of the undersize to the reference size was obtained allowing the calculation of the Work Index with Equation (6).

$$w_i = \frac{44.5}{P_{100}^{0.23} Gbp^{0.82} \left[\frac{10}{\sqrt{P_{80}}} - \frac{10}{\sqrt{F_{80}}} \right]} \tag{6}$$

where

w_i = Bond Work Index (kWh/sht).
P_{100} = test reference size (μm).
Gbp = Grindability Index for the mineral (g/rev.).
F_{80} = grain size corresponding to 80% of the feed undersize (μm).
P_{80} = grain size corresponding to 80% of the final undersize (μm).

Regarding w_i units, although the original Bond formulation uses short tons (sht), results from this work are given in metric tons, after applying the corresponding conversion factor.

3.3. Determination of the Kinetic Parameters

As mentioned before, the kinetic parameters (C and n) were determined in a Bond standard mill with its corresponding ball charge to avoid variability. According to Ersayin et al. [24], the parameters C and n are a function of both the mineral and mill features.

The test feed was a 700 cm^3 representative sample with known particle size distribution. Several grinding runs were performed with different cumulative times (0.5, 1, 2, 3, 4, and 5 min). Size analysis was done between runs, the analyzed sample being reintroduced into the mill before resuming the process. Size analyses were then used first to determine the constant k and then the parameters C and n for each time and size class, using Equations (4) and (5).

3.4. Simulation Based on the Cumulative Kinetic Model

The cumulative kinetic model provides the size distribution of the product from the size distribution of the feed and the grinding times, using Equation (3) and the parameters C and n, determined in the laboratory.

As in the Bond test, an arbitrary initial number of revolutions was set and converted into time units (provided that the mill speed is 70 rpm). Then, using Equation (3), the quantity of product for a reference particle size was calculated. By comparing with the weight of the feed, the resulting undersize and the Gbp were determined for the cycle.

The simulator calculates the new feed (weight and PSD) from the reference size reject of the previous cycle plus the fresh feed that replaces the undersize product of the previous cycle. Next, it calculates the number of revolutions for the following cycle and the grain size distribution of the reconstituted feed. The cycle is repeated until Gbp stability is reached and a circulating load is close to 250%. Finally, the simulation ends after calculating the simulated Work Index value (w_{is}).

A spreadsheet for performing the simulation is available as Supplementary data.

4. Results

4.1. Simulation of the Work Index, Obtaining the Kinetic Parameters through the Conventional Procedure

Table 1 displays the results obtained for the studied usual components of the gangue. It includes the kinetic parameters, the Work Index for the Bond standard procedure, and the Work Index obtained through simulation. The latter two were obtained for a reference size of 100 μm.

Table 1. Comparison between Work Indexes for pure gangue components and reference size 100 μm, obtained through the standard procedure and the simulation (w_i = Work Index; w_{is} = simulated Work Index).

Ore	Kinetic Parameters		R^2	w_{is} (kWh/t)	w_i (kWh/t)	Difference (%)
	C	n				
Feldspar @100 μm	0.000586	1.07	0.98	11.67	12.41	6.0
Limestone @100 μm	0.001789	0.87	0.97	9.66	9.98	3.2
Calcite @100 μm	0.000973	1.09	0.95	6.41	6.30	−1.7
Quartz @100 μm	0.000448	1.07	0.99	13.77	13.88	0.8

Table 2 displays the results obtained for the critical metal ores studied. It includes the kinetic parameters and Work Indexes obtained through Bond standard procedure and simulation (reference size of 100 μm).

Table 2. Comparison between Work Indexes for metal ores and reference mesh of 100 μm, obtained through the standard procedure and the simulation (w_i = Work Index; w_{is} = simulated Work Index).

Ore	Kinetic Parameters		R^2	w_{is} (kWh/t)	w_i (kWh/t)	Difference (%)
	C	n				
M1S1 @100 μm	0.000792	0.89	0.99	19.56	19.25	1.6
M2S1 @100 μm	0.000451	1.03	0.98	15.55	14.83	−4.9
M2S2 @100 μm	0.000486	1.03	0.98	16.61	15.98	3.8
M2S3 @100 μm	0.000303	1.10	0.99	17.06	17.35	−1.7

Bond index values were also determined, with a 150 μm reference size, for feldspar and quartz and then compared with those determined through simulation using CKM. Table 3 compares the results obtained through the standard procedure and the simulation for both minerals. It was noted that the differences between w_i and w_{is} for 150 μm were within the range obtained in the previous cases for a reference size of 100 μm.

Table 3. Work Indexes for feldspar and quartz (reference sizes 100 and 150 µm) obtained through the standard test and the simulation (w_i = Work Index; w_{is} = simulated Work Index).

Ore	Kinetic Parameters		R^2	w_{is} (kWh/t)	w_i (kWh/t)	Difference (%)
	C	n				
Feldspar @150 µm	0.000586	1.07	0.98	11.67	12.41	6.0
Feldspar @100 µm				14.19	14.69	3.4
Quartz @150 µm	0.000448	1.07	0.99	13.77	13.88	0.8
Quartz @100 µm				17.34	17.88	3.0

4.2. Work Index Determination by Obtaining the Kinetic Parameters through the Simplified Procedure

From the results of the tests carried out to determine the kinetic parameters for each sample, it the product size distribution of the longest grinding time test was selected. Then, the slope was determined from the starting time to the end time of the test to obtain k for all the grain size populations involved.

To calculate C and n, the condition was set of using the widest grain size range, with an upper (coarsest) limit of 840 µm and including the finest possible populations without distorting the outcomes significantly. For most cases, this resulted in a lower (finest) limit of 53 µm. Table 4 displays a comparison of the Work Indexes obtained through the Bond standard test and the simulation using the kinetic parameters provided by the simplified procedure. It can be seen that Work Indexes from the simulation using the simplified procedure differ from those from the Bond standard procedure by less than 10%.

Table 4. Compared results between w_i and w_{is} using the simplified procedure to obtain the kinetic parameters.

Ore	Size Interval (µm)		Grinding Time (min)	w_i (kWh/t)		Difference (%)
	Minimum	Maximum		Real	Simulated	
Feldspar @100 µm	53	840	0–4	14.69	13.38	8.91
Feldspar @150 µm	53	840	0–4	12.41	11.20	9.76
Quartz @100 µm	53	840	0–5	17.88	17.82	0.33
Quartz @150 µm	53	840	0–5	13.88	14.63	−5.37
Limestone @100 µm	53	840	0–4	9.98	9.41	5.72
Calcite @100 µm	53	840	0–4	6.30	6.24	0.96
M1S1 @100 µm	74	840	0–3	19.25	21.11	−9.68
M2S1 @100 µm	53	840	0–3	14.83	15.40	−3.85
M2S2 @100 µm	74	840	0–3	15.98	17.01	−6.47
M2S3 @ 100 µm	53	840	0–3	17.35	16.79	3.21

5. Discussion

The simulation of the Work Index test applying the cumulative kinetic model proved to yield results that differ by less than 10% from those from the Bond standard test. This figure agrees with results by Ahmadi and Shahsavari 2009 [25], who proposed a two-step simplified procedure. They applied a simulation based on the cumulative kinetic model and validated their results over three samples of iron ore and one of copper ore. Their results differed by less than 7% from those yielded by the standard procedure. Previously, Aksani and Sönmez 2000 [7] determined the Bond indexes by means of the cumulative kinetic model with values differing by less than 4% with respect those determined with the standard test.

As to determining the kinetic parameters, the possibility of reducing the procedure to only one grinding test with the greatest possible estimated time to obtain the grain size distribution over which the k index and then the parameters C and n could be calculated was evaluated.

As it can be observed in the graph of Figure 1, illustrating the tests performed with quartz, the slope for each particle size class does not vary much over the time span recorded. This suggests that the determination of the constant k would not require obtaining grain size data at intermediate grinding times.

The graph of Figure 2 shows that k values vary for different particle sizes over the grinding time. For grinding times longer than 3 min, the curve is straight for the grain size class between 595 and 840 μm (20 and 30 mesh), and then after a break in slope, a straight line continues again. It can also be noticed that, with grinding time, the different curves adjust and merge into a new slope.

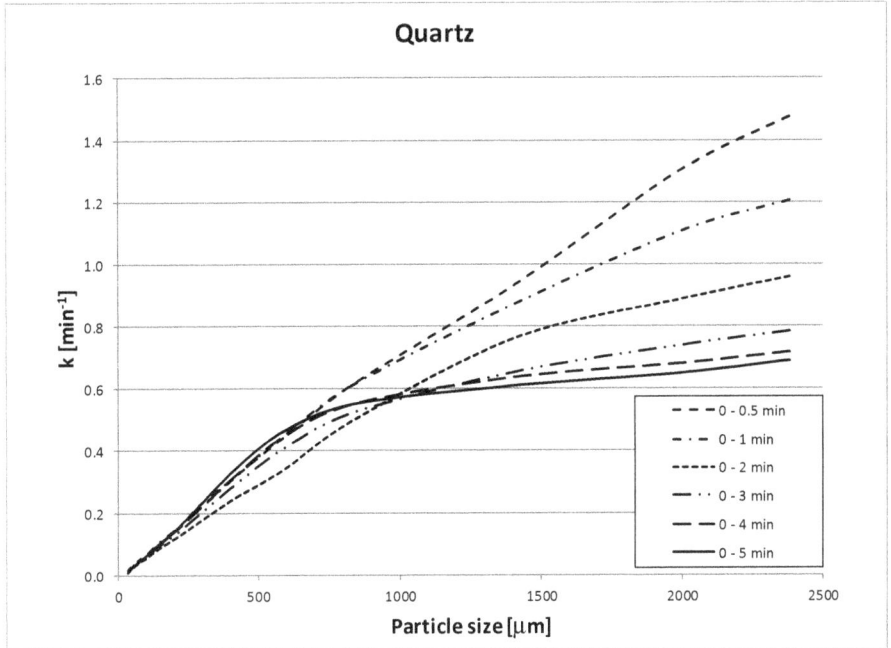

Figure 2. Graph showing the variation of k with particle size for different grinding times, in the case of quartz.

A similar situation occurs for the remaining studied minerals, as the graph of Figure 3 depicts. This graph shows the variation of k with particle size according to grinding times of feldspar. Similarly, Figure 4 displays the results for the ore M2S2.

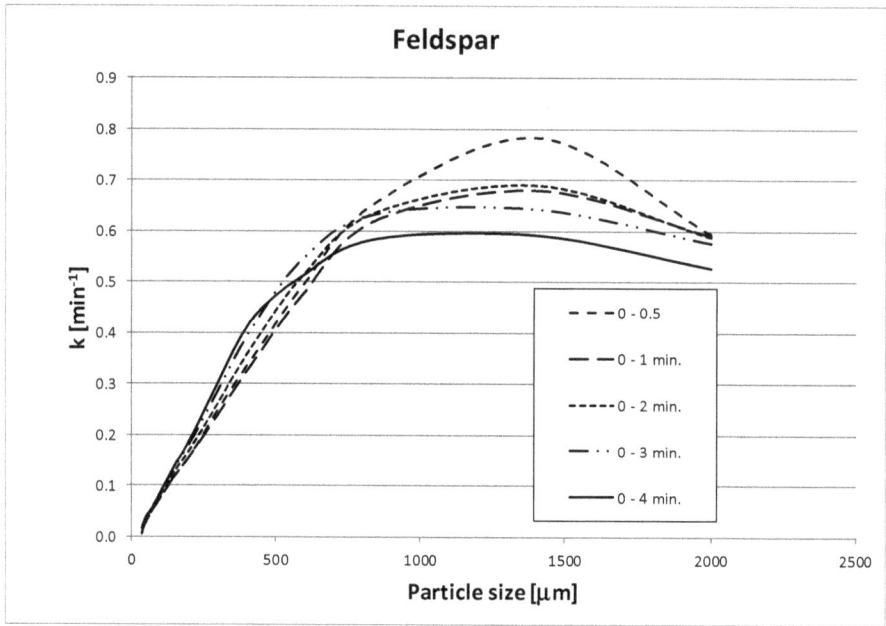

Figure 3. Graph showing the variation of k with particle size for different grinding times, in the case of feldspar.

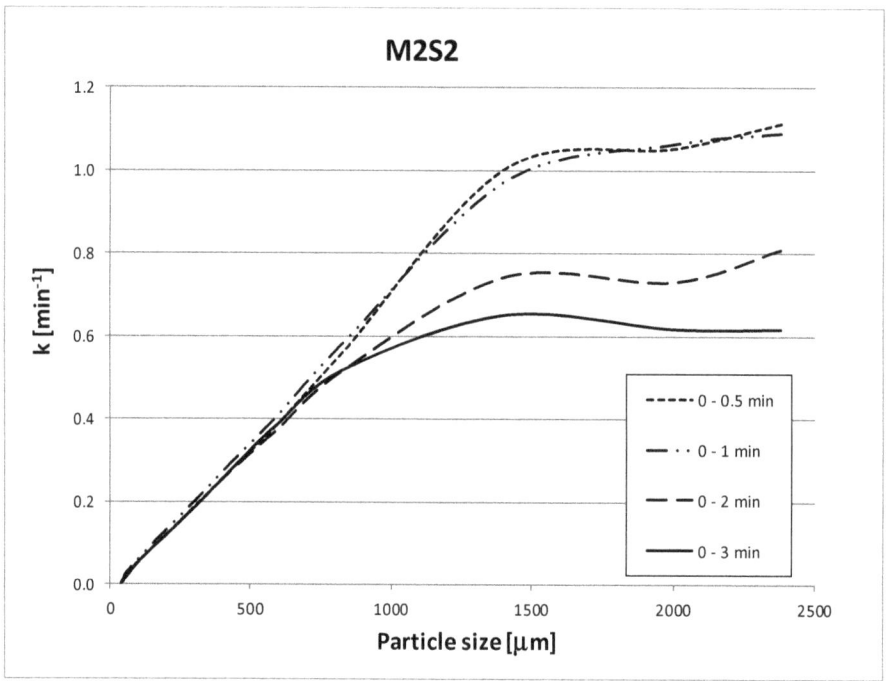

Figure 4. Graph showing the variation of k with particle size for different grinding times, in the case of M2S2 sample.

The graphs show that the curves display a stable shape after a given grinding time. In all cases, a change in slope takes place for particle size coarser than 700 µm. That is the reason why the simplified procedure should not be extended for particle sizes coarser than 840 µm. (20 mesh). This condition does not involve an important limitation since this is the usual particle size range employed in ball mill grinding.

Indeed, it was decided to use the maximum time employed in for each mineral grinding, highlighting the convenience of using a grinding time of at least 5 min to calculate the kinetic parameters with this procedure.

It can be considered that the simplified procedure permits to determine the grinding kinetic parameters C and n, after carrying out one grinding test and two PSD analyses. This can be done rapidly in the laboratory once the sample is prepared. It is considered a valid procedure for a particle size range between 840 and 53 µm with a grinding time of 5 min, using the standard Bond mill with the standard ball charge. The procedure can be summarized as follows:

(1) Feed preparation 100% passing a 3350 µm (6 mesh) sieve followed by gradual crushing to avoid the overproduction of fines.
(2) PSD determination with sieves between 3350 and 37 µm, and determination of the feed characteristic particle size (F_{80}).
(3) Grinding of a quantity equivalent to 700 cm^3 for 5 min.
(4) PSD analysis to determine the product characteristic particle size (P_{80}).
(5) Determination of k for each particle size class (slope from 0 to 5 min).
(6) Determination of the slope (n) and the intercept on the y-axis (C) of the logarithmic-scale graph of k as a function of particle size (k vs. particle size).

Once the kinetic parameters are determined, the Bond test can be simulated, and the PSD can be obtained until reaching the grindability value (Gbp or grams per revolution) for a load of 250% and for the reference size (P_{100}). Once the feed characteristic grain size (F_{80}), the product characteristic grain size (P_{80}), and the grindability index (Gbp) are known, the Work Index can be calculated with Equation (6).

6. Conclusions

The experimental work done and its further analysis permit to draw the following conclusions:

- The conventional cumulative kinetic model (CKM) is a tool that allows determining the Work Index (w_i) for ball mill grinding, simulating the standard procedure of F.C. Bond. The respective results provided, according to literature, differ from less than 7%.
- A simplified procedure has been proposed to obtain the CKM parameters. It is based on determining k with one grinding time since k variation with particle size is rather constant for times less than 5 min. This makes it valid for simulating batch grinding with residence times on the order specified.
- The proposed simplified procedure has been proven to be valid for using the CKM to simulate the F.C. Bond's standard test. It permits to obtain the Work Index in ball mill grinding for a reference size range between 840 (20 mesh) and 53 µm (400 mesh), yielding results that differ by less than 10% with respect to real values. This considerably reduces the involved laboratory work, thus being enough with one grinding run and two PSD determinations.

Supplementary Materials: The following are available online at http://www.mdpi.com/2075-4701/10/7/925/s1, Spreadsheet: Simulation example.

Author Contributions: Conceptualization, J.M.M.A. and V.C.; methodology, J.M.M.A.; software, V.C.; validation, V.C., R.B. and A.T.; formal analysis and investigation, V.C., R.B., A.T., M.P., E.A. and M.P.; resources, V.C.; writing—original draft preparation, V.C.; writing—review and editing, J.M.M.A. and V.C.; supervision, J.M. All authors have read and agreed to the published version of the manuscript.

Funding: This research received no external funding.

Conflicts of Interest: The authors declare no conflict of interest.

References

1. Rittinger von, P.R. *Lehrbuch der Aufbereitungskunde*; Ernst and Korn: Berlin, Germany, 1867.
2. Kick, F. *Das Gesetz der Proportionalen Widerstände und Seine Anwendungen*; Arthur Felix: Leipzig, Germany, 1885.
3. Bond, F.C. Third Theory of Comminution. *Min. Eng. Trans. Aime* **1952**, *193*, 484–494.
4. Bond, F.C. *Crushing and Grinding Calculations*; Allis Chalmers Manufacturing Co.: Milmwaukee, WI, USA, 1961.
5. Jankovic, A.; Dundar, H.; Mehta, R. Relationships between comminution energy and product size for a magnetite ore. *J. South. Afr. Inst. Min. Metall.* **2010**, *110*, 141–146.
6. Sepúlveda, J.; Gutierrez, R. *Dimensionamiento y Optimización de Plantas Concentradoras Mediante Técnicas de Modelación Matemática*; CIMM: Santiago, Chile, 1986.
7. Aksani, B.; Sönmez, B. Simulation of Bond Grindability Test by Using Cumulative Based Kinetic Model. *Miner. Eng.* **2000**, *13*, 673–677. [CrossRef]
8. Coello Velázquez, A.L.; Menéndez-Aguado, J.M.; Brown, R.L. Grindability of lateritic nickel ores in Cuba. *Powder Technol.* **2008**, *182*, 113–115. [CrossRef]
9. Berry, T.F.; Bruce, R.W. A simple method of determining the grindability of ores. *Can. Min. J.* **1966**, *87*, 63–65.
10. Menéndez-Aguado, J.M.; Dzioba, B.R.; Coello-Velazquez, A.L. Determination of work index in a common laboratory mill. *Miner. Metall. Process.* **2005**, *22*, 173–176. [CrossRef]
11. Bonoli, A.; Ciancabilla, F. The Ore Grindability Definition as an Energy Saving Tool in the Mineral Grinding Processes. In Proceedings of the 2nd International Congress "Energy, Environment and Technological Innovation", Rome, Italy, 12–16 October 1992.
12. Chakrabarti, D.M. Simple Approach to Estimation of Work Index. *Trans. Instn. Min. Met. Sect. C Min. Proc. Ext. Met.* **2000**, *109*, 83–89. [CrossRef]
13. Napier- Munn, T.J.; Morrell, S.; Morrison, R.D.; Kojovic, T. *Mineral Comminution Circuits: Their Operation and Optimization*; JKMRC: Queensland, Australia, 1996.
14. Prediction of Bond's work index from measurable rock properties. *Int. J. Miner. Process.* **2016**, *157*, 134–144.
15. Rodríguez, B.Á.; García, G.G.; Coello-Velázquez, A.L.; Menéndez-Aguado, J.M. Product size distribution function influence on interpolation calculations in the Bond ball mill grindability test. *Int. J. Miner. Process.* **2016**, *157*, 16–20. [CrossRef]
16. Lewis, K.A.; Pearl, M.; Tucker, P. Computer Simulation of the Bond Grindability Test. *Miner. Eng.* **1990**, *3*, 199–206. [CrossRef]
17. Silva, M.; Casali, A. Modelling SAG Milling Power and Specific Energy Consumption Including the Feed Percentage of Intermediate Size Particles. *Miner. Eng.* **2015**, *70*, 156–161. [CrossRef]
18. Menéndez-Aguado, J.M.; Coello-Velázquez, A.L.; Dzioba, B.R.; Diaz, M.A.R. Process models for simulation of Bond tests. *Trans. Inst. Min. Metall. Sect. C Miner. Process. Extr. Metall.* **2006**, *115*, 85–90. [CrossRef]
19. Lvov, V.V.; Chitalov, L.S. Comparison of the Different Ways of the Ball Bond Work Index Determining. *Int. J. Mech. Eng. Technol.* **2019**, *10*, 1180–1194. Available online: https://ssrn.com/abstract=3452642 (accessed on 5 June 2020).
20. Gharegheshlagh, H.H. Kinetic grinding test approach to estimate the ball mill work index. *Physicochem. Probl. Miner. Process.* **2016**, *52*, 342–352.
21. Laplante, A.R.; Finch, J.A.; del Villar, R. Simplification of Grinding Equation for Plant Simulation. *Trans. IMM Sec. C* **1987**, *96*, C108–C112.
22. Loveday, B.K. An analysis of comminution kinetics in terms of size distribution parameters. *J. S. Afr. Inst. Min. Metall.* **1967**, *68*, 111–131.
23. Finch, J.A.; Ramirez-Castro, J. Modelling Mineral Size Reduction in Closed-Circuit Ball Mill at the Pine Point Mines Concentrator. *Int. J. Min. Proc.* **1981**, *8*, 67–78. [CrossRef]

24. Ersayin, S.; Sönmez, B.; Ergün, L.; Akasani, B.; Erkal, F. Simulation of the grinding circuit at Gümüşköy Silver Plant, Turkey. *Trans. IMM Sect. C* **1993**, *102*, C32–C38.
25. Ahmadi, R.; Shahsavari, S. Procedure for determination of ball Bond work index in the commercial operations. *Miner. Eng.* **2009**, *22*, 104–106. [CrossRef]

© 2020 by the authors. Licensee MDPI, Basel, Switzerland. This article is an open access article distributed under the terms and conditions of the Creative Commons Attribution (CC BY) license (http://creativecommons.org/licenses/by/4.0/).

 metals

Article

Unveiling the Link between the Third Law of Comminution and the Grinding Kinetics Behaviour of Several Ores

Victor Ciribeni [1], Juan M. Menéndez-Aguado [2,*], Regina Bertero [1], Andrea Tello [1], Enzo Avellá [1], Matías Paez [1] and Alfredo L. Coello-Velázquez [3]

[1] Instituto de Investigaciones Mineras, Universidad Nacional de San Juan, San Juan 5400, Argentina; ciribeni@unsj.edu.ar (V.C.); reginabertero@gmail.com (R.B.); act8383@gmail.com (A.T.); enavella.91@gmail.com (E.A.); matias.p043@gmail.com (M.P.)

[2] Escuela Politécnica de Mieres, University of Oviedo, Gonzalo Gutiérrez Quirós, 33600 Mieres, Spain

[3] CETAM, Universidad de Moa Dr. Antonio Núñez Jiménez, Moa 83300, Cuba; acoello@ismm.edu.cu

* Correspondence: maguado@uniovi.es; Tel.: +34-985458033

Abstract: As a continuation of a previous research work carried out to estimate the Bond work index (w_i) by using a simulator based on the cumulative kinetic model (CKM), a deeper analysis was carried out to determine the link between the kinetic and energy parameters in the case of metalliferous and non-metallic ore samples. The results evidenced a relationship between the CKM kinetic parameter k and the grindability index gbp; and also with the w_i, obtained following the standard procedure. An excellent correlation was obtained in both cases, posing the definition of alternative work index estimation tests with the advantages of more straightforward and quicker laboratory procedures.

Keywords: grinding kinetics; grindability; comminution; bond work index

1. Introduction

The importance of work index determinations in mineral ores comminution operations is without any doubt. The methodology proposed by F.C. Bond [1] is widely used in grinding equipment design and calculations. The crucial point is that it was developed on an enormous data quantity, both at laboratory and industrial scales, yielding sound and reliable results. This fact provided Bond's methodologies with great prestige from its inception and, despite many attempts to develop a technique to replace it over time, it established itself as an essential tool for design and sizing the reduction stages of hundreds of metallurgical plants around the world.

However, the Bond proposal has some shortcomings, pointed out by Gutierrez and Sepulveda [2], Aksani and Sömmez [3] and Menendez Aguado et al. [4], which are summarized below:

- Availability of the standard mill
- Availability of a minimum sample of 10 kg
- Excessive duration of the procedure (in case of some ores)
- Lack of detailed procedure definition (there is no ASTM or ISO specific standard)

These shortcomings have fostered the proposal of alternative grindability characterization procedures. Thus, Lvov and Chitalov [3] performed an in-depth review of several alternative methodologies. Recently, Josefin and Doll [4] proposed an alternative methodology to obtain w_i at a different closure size (P_{100}) than the one tested, and Nikolić and Trumić [5] proposed an alternative procedure when the feed top size (F_{100}) is much lower than 3.35 mm, the top size referenced in the Bond standard methodology (BSM). Moreover, estimating the work index variability from the variability of the geomechanical parameters is the central idea of several alternative procedures, as recently proposed by Park and Kim [6]; this mine-to-mill approach needs further development, but opens a promising way related to mine digitalization strategies for process optimization. Currently, new tools are being developed to estimate w_i, some of which stand out for the reuse of equipment applying modern technologies, simplifying methodologies or applying mathematical models, as it is the case of the following authors:

- Aksani et al. [7] proposed a methodology to obtain the work index by simulation using the CKM and showed results for six different ores, reporting deviations less than 4%.
- Menéndez Aguado et al. [8] showed an alternative methodology based on a non-standard mill, reporting a mean square error of less than 3%.
- Ahmadi et al. [9] presented a methodology with an industrial-scale validation; deviations reported in this case were less than 7%.
- Mwanga et al. [10,11] developed an alternative small sample methodology (300 g) with a geometallurgical approach, reporting mean square error less than 5%.
- Heiskari et al. [12] also presented an alternative methodology using a small sample quantity in a Mergan mill, as an evolution of the former proposal of Niiti [13]. The deviation values reported in this case were less than 4%.

Moreover, Ciribeni et al. [14] proposed a simplified technique to determine the CKM kinetic grinding parameters in order to simulate the Bond test and to validate it by contrasting the results of Au and Ag metalliferous ore samples from various deposits in the Argentine Patagonia. At present, the application of mathematical models to simulate grinding has proven to be a helpful tool for determining the work index, not only in the abovementioned case of Aksani et al. [7], but also in previous work from Lewis et al. [15] and more recently Silva et al. [16]. However, only some authors present alternatives that solve the difficulties of Bond's procedure. This method allows testing with a small sample, especially when looking to obtain the work rate of drill core samples, limiting the sample size to less than a pair of kilograms. This is usually the case of practical geometallurgy, which provides data for the economic and technical evaluation of mineral exploitation and the metallurgical plant, and seeks to predict the mineral behaviour in the metallurgical processes.

The simulator developed by Ciribeni et al. [14] allowed the estimation w_i from the CKM kinetic parameters with a good approximation. Deniz [17] studied the relationship among the Bond standard test parameters and the kinetic parameters following the well-known Austin methodology [18], suggesting several relationships between the grindability index gbp and the set of kinetic data. However, this work was carried out only on one ore, and the practical advantage of this solution is not convenient, given that the determination of Austin parameters can involve even more laboratory work than in the test defined by the BSM in the case of ball mills. Moreover, several papers have been published studying the deviations from the linear kinetic approach [19,20]. However, the CKM procedure provides a quick parameter determination.

Hence, the objective of this research was to study the relationship between the CKM kinetic parameters and the Bond ball mill standard test parameters (gbp, w_i) to propose alternative methodologies of work index estimation with practical advantages. The main hypothesis is that, as suggested by Deniz [17], there can be found a relationship between the CKM kinetic parameter k and the power consumption parameters in the BSM, such as gbp and w_i.

2. Materials and Methods

2.1. The Cumulative Kinetic Model

The cumulative kinetic model is the simple solution defined by Laplante [21], for the equation proposed by Loveday [22] as a first-order kinetic equation. The particle breakage rate in a given size interval is proportional to the mass present in this interval. The kinetic parameter k is defined by the disappearance rate of oversize particles for a given size class (for both batch or continuous grinding—assuming a plug flow regime in the latter one) and can be described with the CKM model as expressed in Equation (1).

$$W_{(x,t)} = W_{(x,0)} \exp(-k \cdot t) \tag{1}$$

wherein:

$W(x,t)$ = cumulative percentage of oversize of size class x at time t.
$W(x,0)$ = cumulative percentage of oversize of size class x in the fresh feed.
k = breakage rate constant (min^{-1})
t = time, (min)

The equation that describes the relationship between the breakage rate and the particle size is shown in Equation (2):

$$k = C \cdot x^n \tag{2}$$

wherein C and n are constants, dependent on the characteristics of the ore and the mill, as described by Ersayin et al. [23]. C and n can be determined experimentally and once known the feed particle size distribution (PSD) the product PSD can be calculated by means of Equation (3).

$$W_{(x,t)} = W_{(x,0)} (\exp(-C \cdot x^n \cdot t)) \tag{3}$$

2.2. Kinetic Parameters Determination

Kinetic parameter k is determined with a small sample in a laboratory mill, as described in [14]. In this case, the standard mill designed by Bond is used to avoid introducing this uncertainty factor when simulating the Bond standard test.

With the same amount of feed from the Bond test (700 cm^3), successive grinding runs are carried out at pre-established time intervals. Once finished each run, a representative sample of the mineral load is taken, and the product PSD is obtained; the sample is returned to the mill, recomposing the load and allowing the performance of the subsequent grinding run.

To simplify this test, the simplified methodology (SIM) presented in [14] proved that a single grinding run could be made to determine the kinetic parameter k, saving time and avoiding excessive manipulation of the sample. The k value is determined for different monosizes, making the linear regression of the cumulative retained for the final milling time, using Equation (4):

$$\mathrm{Ln}\left(W_{(x,t)}\right) - \mathrm{Ln}\left(W_{(x,0)}\right) = kt \tag{4}$$

2.3. Experimental Procedure

2.3.1. Sample Preparation

For this work, samples from three metalliferous ores from Argentine Patagonia were selected and prepared according to the conventional preparation scheme used to prepare the feed in a Bond's ball mill standard test (Figure 1). In each case, the sample amount prepared was enough to carry out the SIM tests to obtain the kinetic parameters k [14] and also to obtain the BSM work index [1], which will be used for validation purposes.

PSD were obtained in a Ro-Tap sieve with sieves 203 mm in diameter (ASTM certified). Sampling was carried out in a Sieving Riffler Quantachrome eight sector rotary sampler.

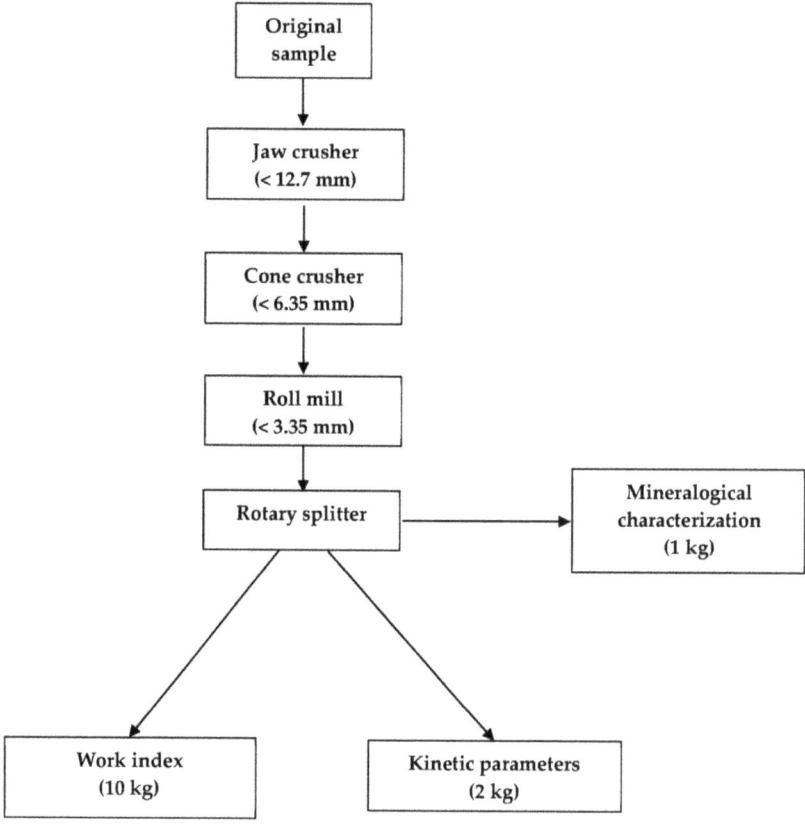

Figure 1. Samples preparation flowsheet.

2.3.2. Sample Characterization

To perform this study, data from Ciribeni et al. [14] were used. However, some additional ores were considered in this research and are included below.

The new samples came from metalliferous deposits (Au, Ag) from the metallogenetic province "Macizo del Deseado" (Argentine Patagonia):

- Low sulphidation (LS) ore: several samples were taken from this ore, which comes from a low sulphidation hydrothermal deposit, formed mainly by veins of silica in the form of quartz, chalcedony and opal; native gold is present, and silver can be found in a wide range of minerals (electrum, sulfosalts, cassiterite, galena, pyrite and chalcopyrite, among other minerals).
- High sulphidation (HS) ore: comes from deposits of the epithermal type of medium sulphidation, which is made up of quartz, carbonates and to a lesser extent Au, and Ag sulphides and sulphosalts, in addition to Pb, Cu and Zn.

2.3.3. Determination of Kinetic Parameters and Work Index

The kinetic parameter k was determined following the SIM methodology [14]. It is carried out in a Bond standard ball mill, with sample feed 700 cm^3 (prepared according to the procedure presented in Figure 1). After PSD feed determination, a sole 5 min grinding run is performed, and the product PSD is obtained. This grinding time value was selected considering that grinding runs in the Bond standard test do not usually exceed 350 revolutions (5 min, at 70 rpm). For each size interval, k is calculated (Equation (4)).

Using Equation (2) for that set of k and x values, C and n for each ore are calculated, and an estimation of work index by CKM simulation, $w_{i,s}$, is performed [14].

The Bond work index (w_i) was determined following BSM, the standard methodology developed by F. C. Bond [1]. The ball mill work index laboratory test is conducted by grinding an ore sample prepared to 100% passing 3.36 mm to product size in the range of 45–150 µm, thus determining the ball mill w_i. Several sources of variability, mainly due to a lack of procedure definition were identified by García et al. [24]. With the aim of reducing that variability, a detailed description of the test can be found in the proposal of the Global Mining Guidelines Group [25].

3. Results and Discussion

3.1. Work Index Calculation and Estimation

Table 1 shows the actual BSM w_i values obtained versus the work index estimation by CKM simulation ($w_{i,s}$). LS-CMLM1 and LS-CMVM1 samples were tested at a reference size P_{100} = 74 µm; estimated values by simulation differ less than 4%. Meanwhile, HS-CCTUM1 sample was tested at P_{100} = 149 µm, and the estimation difference with the actual w_i value was rounded by 6%.

Table 1. Comparison between w_i and $w_{i,s}$ for different metalliferous ores.

Sample	P_{100}	w_i [kWh/t]	$w_{i,s}$ [kWh/t]	Difference [%]
LS-CMLM1	74	26.59	27.63	−3.91
LS-CMVM1	74	25.17	24.56	2.42
HS-CCTUM1	149	13.82	12.96	6.22

Results of w_i and $w_{i,s}$ calculations in the considered samples from previous research [9] complete the subsequent Tables 2 and 3.

Table 2. Grindability and kinetic parameters obtained experimentally.

Sample	P_{100} [µm]	F_{80} [µm]	P_{80} [µm]	w_i [kWh/t]	gbp [g/rev]	k (@ P_{100})
LS-CMLM1	74	1938	54	26.59	0.8038	0.03667
LS-CMVM1	74	2053	53	25.17	0.6907	0.03656
HS-CCTUM1	149	2469	96	13.82	1.3427	0.12543
LS-CNM1	149	2508	113	21.17	1.0176	0.06539
HS-CVM1	149	2284	115	16.31	1.4784	0.09132
LS-CVM2	149	2432	116	17.57	1.2500	0.08693
LS-CVM3	149	2333	114	19.08	1.1180	0.08143
Quartz	149	2572	117	15.27	1.5773	0.10141
Quartz	105	2552	83	19.89	1.1048	0.06492
Feldspar	149	1841	115	13.65	1.9112	0.13860
Feldspar	105	1676	81	16.16	1.3567	0.09183
Limestone	149	2407	108	10.88	2.2533	0.15128
Calcite	149	2497	112	6.93	3.9221	0.25711
Cryst. limestone	149	2062	112	8.46	3.3308	0.20919
Cryst. limestone	105	1926	79	10.91	2.2591	0.14956

Table 3. Comparison w_i versus $w_{i,e1}$.

Sample	P_{100} [μm]	w_i [kWh/t]	gbp [g/rev]	$w_{i,e1}$ [kWh/t]	gbp_e [g/rev]	Difference [%]
LS-CMLM1	74	26.59	0.8038	24.84	0.5869	6.59
LS-CMVM1	74	25.17	0.6907	24.48	0.5852	2.73
HS-CCTUM1	149	13.82	1.3427	11.23	1.8863	18.73
LS-CNM1	149	21.17	1.0176	20.77	1.0073	1.88
HS-CVM1	149	16.31	1.4784	16.37	1.3869	−0.39
LS-CVM2	149	17.57	1.2500	16.97	1.3226	3.44
LS-CVM3	149	19.08	1.1180	17.77	1.2422	6.88
Quartz	149	15.27	1.5773	15.09	1.5211	1.15
Quartz	105	19.89	1.1048	18.65	1.0003	6.24
Feldspar	149	13.65	1.9112	12.15	2.0790	11.00
Feldspar	105	16.16	1.3567	14.74	1.3944	8.78
Limestone	149	10.88	2.2533	10.44	2.2647	4.00
Calcite	149	6.93	3.9221	6.94	3.8140	−0.10
Cryst. limestone	149	8.46	3.3308	8.42	3.1125	0.45
Cryst. limestone	105	10.91	2.2591	9.66	2.2395	11.48

3.2. Relationships between Grindability and Kinetic Constant k

Table 2 summarises the work indices determined by BSM and estimated by CKM with kinetic indices k determined by the SIM procedure. The considered metalliferous ore samples came from the current research tests and the former research ones. Some non-metallic minerals from the former research are also included. All data are used to unveil the links between BSM parameters and those obtained by grinding kinetics (SIM).

After plotting gbp (determined with the standard procedure) versus the kinetic parameter k (calculated by the SIM methodology), as is shown in Figure 2, a linear estimation can be obtained (Equation (5)) with a correlation coefficient of 95.8%.

$$gbp = 14.97 \cdot k \qquad (5)$$

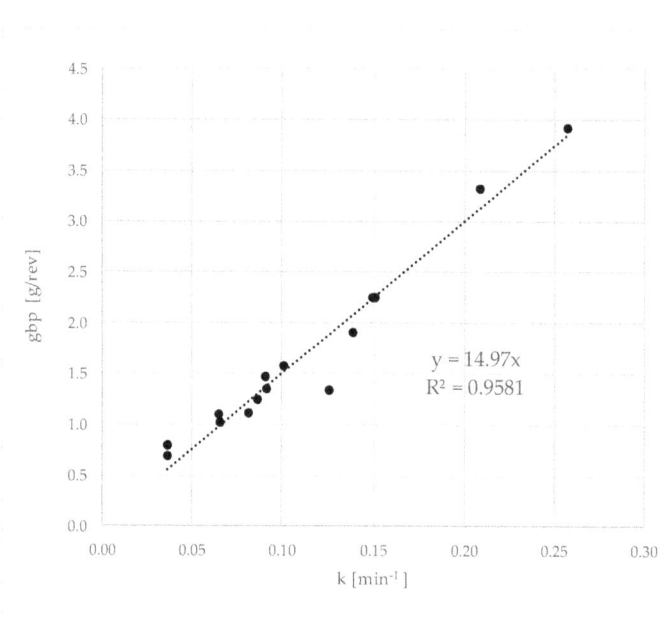

Figure 2. Plot gbp (BSM) versus k (SIM).

According to the BSM procedure, w_i can be calculated for each P_{100} once gbp, F_{80} and P_{80} are known. A new work index estimation, $w_{i,e1}$, can be suggested considering the gbp estimated value (gbp_e) in Equation (5), and F_{80} and P_{80} estimated by the SIM procedure, in each case. Table 3 shows the results obtained from this new estimation proposal, wherein work index differences are in general less than 10% for each ore. However, there are three values above 10% and one reaching 18%.

Figure 3 depicts the relationship of w_i versus $w_{i,e1}$; a linear correlation (Equation (6)) can be plotted, with a correlation coefficient of 98.22%.

$$w_i = 0.962 \cdot w_{i,e1} - 0.28 \tag{6}$$

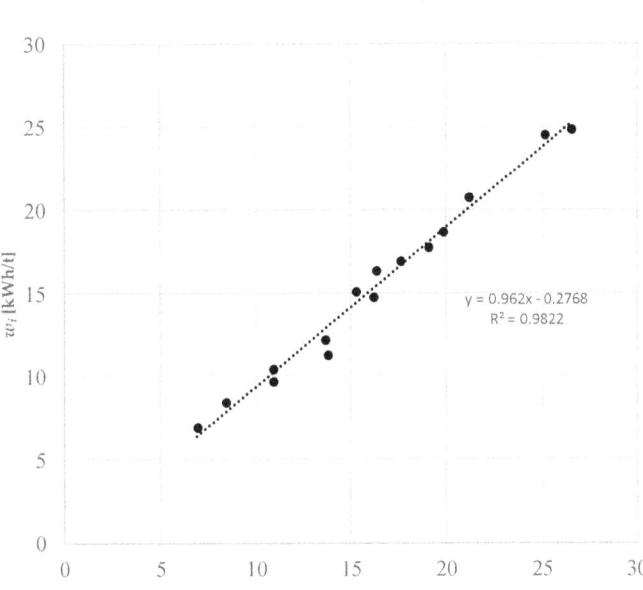

Figure 3. Linear correlation between w_i and $w_{i,e1}$.

3.3. Linking the Work Index w_i and the Kinetic Constant k

As a consequence of the relationships evidenced in Equations (5) and (6), it can be inferred that there should be a correlation between the Bond work index and the kinetic constant k at each monosize. Figure 4 depicts this relationship between w_i and k from the actual data gathered in Table 2, wherein a logarithmic correlation (Equation (7)) poses a correlation coefficient of 98.37%.

$$w_i = -10.07 \cdot \ln(k) - 7.28 \tag{7}$$

Table 4 shows a comparison of actual work index values versus work index estimation using Equation (7); in all cases, differences are lower than 9%.

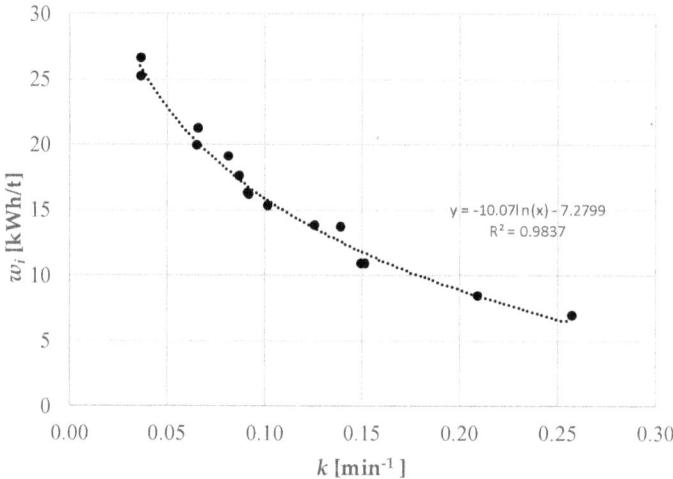

Figure 4. Logarithmic correlation w_i versus k.

Table 4. Comparison between w_i and work index estimation using Equation (7), $w_{i,e2}$.

Sample	P_{100} [μm]	w_i [kWh/t]	k [@P_{100}]	$w_{i,e2}$ [kWh/t]	Difference [%]
LS-CMLM1	74	26.59	0.03667	26.09	1.89
LS-CMVM1	74	25.17	0.03656	26.12	−3.77
HS-CCTUM1	149	13.82	0.12543	13.67	1.10
LS-CNM1	149	21.17	0.06539	20.25	4.36
HS-CVM1	149	16.31	0.09132	16.87	−3.45
LS-CVM2	149	17.57	0.08693	17.37	1.13
LS-CVM3	149	19.08	0.08143	18.03	5.50
Quartz	149	15.27	0.10141	15.81	−3.57
Quartz	105	19.89	0.06492	20.32	−2.16
Feldspar	149	13.65	0.13860	12.66	7.26
Feldspar	105	16.16	0.09183	16.82	−4.06
Limestone	149	10.88	0.15128	11.78	−8.23
Calcite	149	6.93	0.25711	6.42	7.38
Cryst. limestone	149	8.46	0.20919	8.50	−0.49
Cryst. limestone	105	10.91	0.14956	11.89	−8.99

In Figure 5, the estimated work index using Equation (7) $w_{i,e2}$ is plotted versus w_i, revealing the unexpected linear correlation shown in Equation (8), with a correlation coefficient of 98.37%

$$w_i = 0.997 \cdot w_{i,e2} - 0.001 \tag{8}$$

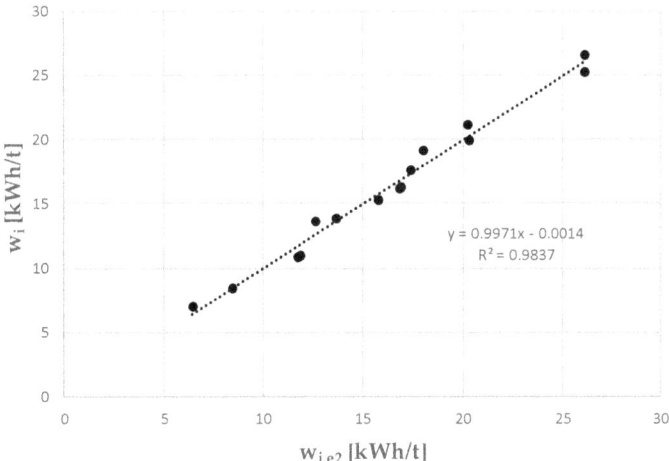

Figure 5. Linear correlation w_i versus $w_{i,e2}$.

3.4. Discussion

Table 1 presented the work index estimation by CKM simulation ($w_{i,s}$) versus the actual w_i values, in the case of three different samples. Results show deviations less than 6%. This procedure saves laboratory time (8 to 2 h, approximately) and reduces sample needs from 10 kg to less than 1.5 kg.

Intending to get a more significant reduction, the research work focused on searching for a correlation between the k kinetic parameter, determined by the SIM methodology developed by Ciribeni et al. [10], and the grindability index, gbp. As can be observed in Figure 2, a good linear correlation (Equation (5)) was obtained, opening the possibility of estimating gbp from the k value (which can be obtained in the laboratory more easily and quickly) and thus providing a new proposal of work index estimation, $w_{i,e1}$.

In Table 3, the comparison between w_i (Bond work index value obtained following the BSM) and $w_{i,e1}$ showed differences, in general, lower than 9%, being in some cases greater than 10%, even reaching 18% in one specific case. This fact can raise the consideration that this methodology is a bit erratic.

On the other hand, the study of the relationship between w_i and k presents an adjustment to a logarithmic function with a correlation coefficient higher than 98%, which is more than acceptable considering that this result was obtained adjusting data of fifteen actual w_i determinations on different ores and with different P_{100}. The estimation of w_i using Equation (7) posed differences lower than 9% in all cases and for all reference sizes. Moreover, in Table 4, it is observed that in the case of metalliferous ores, differences are below 5.50%, with higher variability in the case of non-metalliferous ones. Amadi and Shahsavari [9] reported deviations lower than 7%, and Aksani and Sönmez reported values lower than 4%, using in both cases the CKM simulation-based methodology, that is, in the same order of magnitude.

According to results depicted in Figure 5, where the linear fitting (Equation (8)) casts a correlation coefficient higher than 98% with a slope very close to one and an almost zero intercept, it seems feasible and accurate enough to perform the work index estimation at a given P_{100}, just by knowing the kinetic parameter k obtained at P_{100} and by using Equation (7).

The combination of the SIM methodology [14] with the correlation of k and w_i provides a quick solution, with a minimum sample amount needed, in order to estimate the work index. An additional advantage is the reduction of procedures involving sample manipulation and quartering at the lab, which are usual sources of experimental error.

4. Conclusions

From the results obtained in this research, the following conclusions can be highlighted:

- Once the CKM kinetic parameter k for the given reference sieve P_{100} was known, it was possible to estimate the BSM ball mill work index at that reference size, with differences lower than 9% with the Bond standard methodology.
- It was found that a linear fit yielded a correlation coefficient higher than 96% between gbp and the kinetic parameter k (Equation (5)). The line has slope fifteen and zero intercept. However, estimating w_i by determining gbp with Equation (5) and calculating w_i with the Bond equation gives some erratic values.
- With fifteen different ore samples and for three different P_{100}, a logarithmic correlation w_i versus k was obtained (Equation (7)) with a correlation coefficient higher than 98%. It can be suggested that the logarithmic function in Equation (7) could be a valuable tool as a quick alternative to Bond's standard test in the day-by-day grindability control.
- The comparison between w_i and $w_{i,e2}$ (Equation (8)) shows a linear fit whose slope is unity and the ordinate to the origin is negligible, with a correlation coefficient higher than 98%.
- The use of k versus w_i correlation provides a quick solution, with a minimum sample amount need, in order to estimate the work index.

Author Contributions: Conceptualization, J.M.M.-A., A.L.C.-V. and V.C.; methodology, J.M.M.-A. and V.C.; software, V.C.; validation, V.C., R.B. and A.T.; formal analysis and investigation, V.C., R.B., A.T., M.P., E.A. and M.P.; resources, J.M.M.-A. and V.C.; writing—original draft preparation, V.C.; writing—review and editing, J.M.M.-A. and A.L.C.-V.; supervision, A.L.C.-V. All authors have read and agreed to the published version of the manuscript.

Funding: This research does not receive external funding.

Institutional Review Board Statement: Not applicable.

Informed Consent Statement: Not applicable.

Data Availability Statement: Not applicable.

Conflicts of Interest: The authors declare no conflict of interest.

References

1. Bond, F.C. Crushing and grinding calculations. Part I. *Br. Chem. Eng.* **1961**, *6*, 378–385.
2. Gutiérrez, L.; Sepúlveda, J. *Dimensionamiento y Optimización de Plantas Concentradoras, Mediante Técnicas de Modelación Matemática*; CIMM: Santiago, Chile, 1986.
3. Valerevich Lvov, V.; Sergeevich Chitalov, L. Comparison of the Different Ways of the Ball Bond Work Index Determining. *Int. J. Mech. Eng. Technol.* **2019**, *10*, 1180–1194. Available online: https://ssrn.com/abstract=3452642 (accessed on 2 July 2021).
4. Nikolić, V.; Trumić, M. A new approach to the calculation of Bond work index for finer samples. *Miner. Eng.* **2021**, *165*, 106858. [CrossRef]
5. Josefin, Y.; Doll, A.G. Correction of Bond Ball Mill Work Index Test for Closing Mesh Sizes. Procemin-Geomet 2018. In Proceedings of the 14th International Mineral Processing Conference & 5th International Seminar on Geometallurgy, Santiago, Chile, 28–30 November 2018; pp. 1–12.
6. Park, J.; Kim, K. Use of drilling performance to improve rock-breakage efficiencies: A part of mine-to-mill optimization studies in a hard-rock mine. *Int. J. Min. Sci. Technol.* **2020**, *30*, 179–188. [CrossRef]
7. Aksani, B.; Sönmez, B. Simulation of Bond grindability test by using cumulative based kinetic model. *Miner. Eng.* **2000**, *13*, 673–677. [CrossRef]
8. Menéndez-Aguado, J.M.; Dzioba, B.R.; Coello-Velazquez, A.L. Determination of work index in a common laboratory mill. *Miner. Metall. Process.* **2005**, *22*, 173–176. [CrossRef]
9. Ahmadi, R.; Shahsavari, S. Procedure for determination of ball Bond work index in the commercial operations. *Miner. Eng.* **2009**, *22*, 104–106. [CrossRef]
10. Mwanga, A.; Rosenkranz, J.; Lamberg, P. Testing of ore comminution behavior in the geometallurgical context: A review. *Minerals* **2015**, *5*, 276–297. [CrossRef]
11. Mwanga, A.; Rosenkranz, J.; Lamberg, P. Development and experimental validation of the Geometallugical Comminution test (GCT). *Miner. Eng.* **2017**, *108*, 109–114. [CrossRef]

12. Heiskari, H.; Kurki, P.; Luukkanen, S.; Gonzalez, M.; Lehto, H.; Liipo, J. Evelopment of a comminution test method for small ore samples. *Miner. Eng.* **2019**, *130*, 5–11. [CrossRef]
13. Niitti, T. Rapid evaluation of grindability by a simple batch test. In Proceedings of the International Mineral Processing Congress Proceedings, Prague, Czech Republic, 1–6 June 1970; pp. 41–46.
14. Ciribeni, V.; Bertero, R.; Tello, A.; Puertas, M.; Avellá, E.; Paez, M.; Menéndez Aguado, J.M. Application of the Cumulative Kinetic Model in the Comminution of Critical Metal Ores. *Metals* **2020**, *10*, 925. [CrossRef]
15. Lewis, K.A.; Pearl, M.; Tucker, P. Computer Simulation of the Bond Grindability Test. *Miner. Eng.* **1990**, *3*, 199–206. [CrossRef]
16. Silva, M.; Casali, A. Modelling SAG Milling Power and Specific Energy Consumption Including the Feed Percentage of Intermediate Size Particles. *Miner. Eng.* **2015**, *70*, 156–161. [CrossRef]
17. Deniz, V. Relationships Between Bond's Grindability (Gbg) and Breakage Parameters of Grinding Kinetic on Limestone. *Powder Technol.* **2004**, *139*, 208–213. [CrossRef]
18. Austin, L.G.; Klimpel, R.R.; Luckie, P.T. *Process Engineering of Size Reduction: Ball Milling*; SME-AIME: New York, NY, USA, 1984.
19. Bilgili, E.; Scarlett, B. Population balance modeling of nonlinear effects in milling processes. *Powder Technol.* **2005**, *153*, 59–71. [CrossRef]
20. Petrakis, E.; Komnitsas, K. Development of a Non-linear Framework for the Prediction of the Particle Size Distribution of the Grinding Products. *Min. Metall. Explor.* **2021**, *38*, 1253–1266.
21. Laplante, A.R.; Finch, J.A.; del Villar, R. Simplification of Grinding Equation for Plant Simulation. *Trans. Inst. Min. Metall. (Sect. C)* **1987**, *96*, C108–C112.
22. Loveday, B.K. An analysis of comminution kinetics in terms of size distribution parameters. *J. S. Afr. Inst. Min. Metall.* **1967**, *68*, 111–131.
23. Ersayin, S.; Sönmez, B.; Ergün, L.; Aksani, B.; Erkal, F. Simulation of the Grinding Circuit at Gümüşköy Silver Plant, Turkey. *Trans. Inst. Min. Metall. (Sect. C)* **1993**, *102*, C32–C38.
24. García, G.G.; Oliva, J.; Guasch, E.; Anticoi, H.; Coello-Velázquez, A.L.; Menéndez-Aguado, J.M. Variability Study of Bond Work Index and Grindability Index on Various Critical Metal Ores. *Metals* **2021**, *11*, 970. [CrossRef]
25. GMG—Global Mining Guidelines Group. Determining the Bond Efficiency of Industrial Grinding Circuits. 2016. Available online: https://gmggroup.org/wp-content/uploads/2016/02/Guidelines_Bond-Efficiency-REV-2018.pdf (accessed on 1 July 2021).

Article

Kinetics of Dry-Batch Grinding in a Laboratory-Scale Ball Mill of Sn–Ta–Nb Minerals from the Penouta Mine (Spain)

Jenniree V. Nava [1], Teresa Llorens [1] and Juan María Menéndez-Aguado [2,*]

1. Strategic Minerals Spain, S.L. Ctra. OU-0901 Km 14, Penouta Mine, Viana do Bolo, 32558 Ourense, Galicia, Spain; jvnava@strategicminerals.com (J.V.N.); tllorens@strategicminerals.com (T.L.)
2. Escuela Politécnica de Mieres, University of Oviedo, Gonzalo Gutiérrez Quirós, 33600 Mieres, Asturias, Spain
* Correspondence: maguado@uniovi.es; Tel.:+34-985458033

Received: 17 November 2020; Accepted: 16 December 2020; Published: 17 December 2020

Abstract: The optimization of processing plants is one of the main concerns in the mining industry, since the comminution stage, a fundamental operation, accounts for up to 70% of total energy consumption. The aim of this study was to determine the effects that ball size and mill speed exert on the milling kinetics over a wide range of particle sizes. This was done through dry milling and batch grinding tests performed on two samples from the Penouta Sn–Ta–Nb mine (Galicia, Spain), and following Austin methodology. In addition, the relationships amongst Sn, Ta and Nb content, as metals of interest, the specific rate of breakage S_i, the kinetic parameters, and the operational conditions were studied through X-Ray fluorescence (XRF) techniques. The results show that, overall, the specific rate of breakage S_i decreases with decreasing feed particle size and increasing ball size for most of the tested conditions. A selection function, α_T, was formulated on the basis of the ball size for both Penouta mine samples. Finally, it was found that there does exist a direct relationship amongst Sn, Ta and Nb content, as metals of interest, in the milling product, the specific rate of breakage S_i and the operational–mineralogical variables of ball size, mill speed and feed particle size.

Keywords: ball mill; kinetic grinding; specific grinding rate; Sn–Ta–Nb; Penouta Mine

1. Introduction

In the mining industry, the comminution stage can represent up to 70% of the energy consumed in a mineral processing plant [1–5]. With ball-mill grinding being one of the most energy-consuming techniques, setting the optimal values of the operational and mineralogical parameters for efficient grinding is a key target in mineral processing plants [6–10]. Ball size is one of the key factors of ball-mill efficiency [11,12], and may have a significant financial impact [13]. The population balance model (PBM) has been widely used in ball mills [14]. This model is a simple mass balance to reduce size being used in fragmentation models [15]. Several methods have been implemented to determine those functions. Some were based on simple laboratory-scale grinding essays [16–21], whereas others were based on industry-scale works [22–26]. This paper focuses on studying the specific rate of breakage S_i and its kinetic parameters based on the Austin methodology [27], which assumes that the specific rate of breakage (S_i) is a constant of proportionality that may or may not behave as a first-order function, whereas the function of fracture (B_{ij}) does not change with grinding time.

Tantalum and Niobium are considered critical raw material in the EU, due to their features and applications in a wide range of industrial sectors, and the strong EU import dependence [28]. This makes it of paramount importance to increase the research in the mineral deposits that contain them, and to optimize the processing plants to increase their efficiency and to minimize their energy consumption.

One of those processing plants lies in the Penouta Sn–Ta–Nb mine. Currently, it is the only working mine in Europe producing Ta and Nb concentrates as its main product. This is done by reprocessing the tailing ponds generated by the mining works up to the 1980s, and it is pending authorizations to start mining the source rock. Due to that, two types of sample have been studied: (i) unaltered rock from the Sn-, Ta- and Nb-enriched albite leucogranite (Bedrock); and (ii) material from the tailing ponds (Tailings Pond).

The aim of this work was to study the effects of ball size on milling kinetics, operating at different mill speeds and with a wide range of feed particle size. This was done through dry milling and batch grinding tests, following the methodology proposed by Austin et al. [7] and developed in [9,29]. In addition, it studied the relationships amongst the evolution of Sn, Ta and Nb content, as metals of interest, determined by XRF, the specific rate of breakage S_i, and the operational conditions for both samples, Bedrock and Tailings Pond, from the Penouta mine.

2. Theoretical Background

The population balance model (PBM) has been widely used in ball mills. This model is based on determining the particle size distribution grouped in size classes. A mass balance for the class i in a well-mixed grinding process is done by means of Equation (1), where comminution is linear, and a first-order kinetic fragmentation is assumed [19].

$$\frac{dw_i}{dt} = -S_i w_i(t) + \sum_{j=1}^{i-1} b_{ij} k_j w_j(t) \tag{1}$$

where $w_i(t)$ is the particle mass fraction of size class i at grinding time t. The first term of the right-hand side is the mass fraction of particles of the monosize i that break and, thus, no longer belong to that monosize. S_i is the specific rate of breakage. The second term represents the contribution of all monosizes coarser than i that at breaking produce particles of monosize i. The fracture rate or fracture velocity of a monosize material can be expressed by Equation (2):

$$\frac{-dw_i}{dt} = S_i w_i(t) \tag{2}$$

where S_i is a constant of proportionality called the specific rate of breakage or probability of fracture, whose unit is t^{-1}. Assuming that S_i does not change with time, the integral results in Equation (3).

$$\log(w_i(t)) - \log(w_i(0)) = \frac{-S_i(t)}{2.3} \tag{3}$$

where $w_i(t)$ and $w_i(0)$ are the mass fractions for size class i, at grinding times t and 0, respectively. S_i is the specific rate of breakage. Following the methodology proposed by Austin et al. [7] once S_i values have been obtained through slope determination, they are plotted to the particle size, and Equation (4) is proposed to study the behavior of the specific rate of breakage S_i.

$$S_i = \alpha_T \cdot X_i^\alpha \cdot Q_i \tag{4}$$

where X_i is the upper size limit of the interval (in mm), and α_T, is a parameter that depends on milling conditions and, is the breakage rate for size $x_i = 1$ mm, while α is a characteristic parameter depending on material properties; Q_i is a correction factor, which is 1 for small particles (normal breakage, which was assumed in this case) and less than 1 for large particles that need to be nipped and fractured by the grinding media (abnormal breakage); S_i increases up to a specified size x_m (optimum feed size), but above this size breakage rates decrease sharply [9].

Rotating critical speed of the mill, Nc, is calculated with Equation (5).

$$Nc = \frac{42.3}{\sqrt{D-d}} \qquad (5)$$

where D is the mill diameter and d is the ball diameter (in m). Ball mill filling volume is calculated using Equation (6), assuming that the bed porosity of balls is 40%.

$$J = \left(\frac{mass\ of\ balls}{ball\ density \times mill\ volume}\right) \times \frac{1.0}{0.6} \qquad (6)$$

On the other hand, Austin and Brame [25] calculated the selection function α_T in a general way through Equation (7).

$$\alpha_T = \frac{v_c - 0.1}{1 + e^{[15.7(v_c - 0.94)]}} \qquad (7)$$

where v_c is the mill speed expressed as the fraction of critical speed.

3. Methodology

3.1. Sample Characterization

First, a representative sample of a metric tonne from each of both areas of interest of the Penouta mine, Bedrock and Tailings Pond, was crushed at a size of −4 mm using a jaw crusher. Working samples were obtained after homogenization and quartering using a Jones splitter. Feed monosizes of 3350/2000, 2000/1000, 1000/500, 500/250, 250/125, 125/75 and 75/45 µm were obtained in a sieve shaker using a series of sieves with the openings of above.

Next, feed was characterized by means of grain-size analysis of the above-mentioned size fractions and by means of XRF analysis of fused bead samples using a 4 kW BRUKER spectrometer (Leipzig, Germany), specifically calibrated for this mineralogy, installed in the ALS laboratory at the Penouta mine.

3.2. Calculation of the Critical Speed and Initial Conditions for the Grinding Kinetics Tests

Critical speed was calculated using Equation (5). Table 1 displays mill rotational speeds as a function of ball monosizes for each test.

Table 1. Working speeds for the grinding kinetic tests.

d, Balls Size (mm)	Nc, Mill Critical Speed (rpm)	Work Speed/75% Nc (rpm)	Work Speed/85% Nc (rpm)
19.1	105.9	79.4	90.0
22.3	106.9	80.2	90.1
31.8	110.3	82.8	93.8

Dry batch milling kinetics tests were done in a lab-scale mill, 17.8 cm in diameter and 4.5 L in capacity, on a 600 cm^3 representative volume of each Penouta mine sample. The mill charge consisted of 5.0 kg of steel balls, of 19.0 mm, 22.0 mm and 31.0 mm monosizes. Fill fraction was calculated from Equation (6). Seven feed size fractions (3350/2000, 2000/1000, 1000/500, 500/250, 250/125, 125/75, 75/45 µm) were used to evaluate the influence of this mineralogical variable in the kinetic parameters. Mill discharges were marked through 5 grinding times (0.5; 1; 1.5; 3.5; 7.5 min). This way, each sample was dumped from the mill, and then it went through a grain size analysis by means of dry sieving. In addition, after completing the grinding time, Sn, Ta and Nb content was determined for the undersize to grid i in order to evaluate the evolution of Sn, Ta and Nb grades, with respect to the specific rate of breakage.

3.3. Determination of the Specific Rate of Breakage (S_i) and the Kinetic Parameters (α_T, α)

Following the BII methodology introduced in [27], after measuring the oversize weight for each grinding time, the graph log ($w_i(t)/w_i(0)$) vs. time is plotted for each monosize. The equation of each curve and thus the S_i value are obtained through linear fitting using Equation (3). Then, the S_i values for each monosize are plotted and using Equation (4) the parameters (α_T and α) are calculated for each condition of mill speed and ball size. This allows studying the influence of these two operational variables on the specific rate of breakage and the kinetic parameters α_T and α. The selection function α_T was formulated by means of Equation (9). Nevertheless, this is a general equation, so a specific formula was generated to characterize the samples Bedrock and Tailings Pond from Penouta mine.

4. Results and Discussion

4.1. Chemical Characterisation of the Feed

Both the Tailings Pond and Bedrock head samples display the grain size distribution shown in Figure 1.

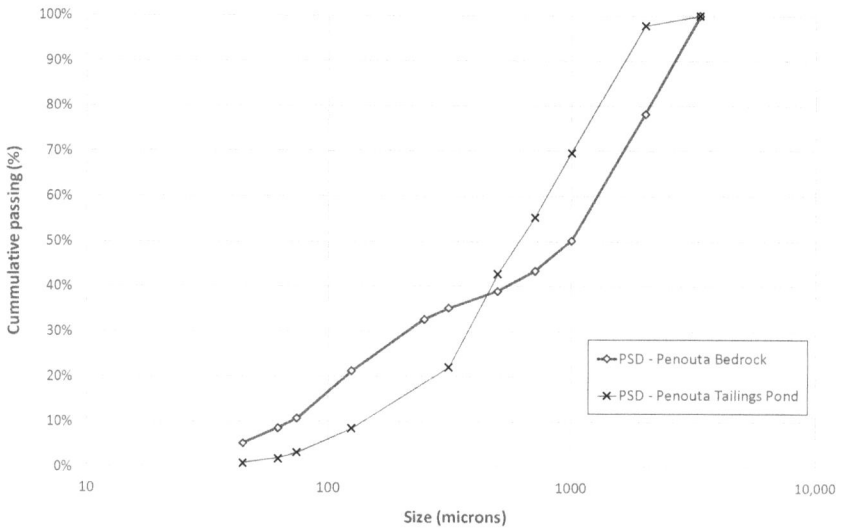

Figure 1. Grain size distribution curves for the Penouta mine head samples.

Bedrock and Tailings Pond samples display an F_{80} of 2110 µm and 1369 µm, respectively. The smaller F_{80} value of Tailings Pond sample results from this material having been previously processed during the mining activities throughout the 20th century, until the 1980s.

The representative chemical composition for both the Tailings Pond and Bedrock head samples is shown in Table 2 and has been obtained through XRF analysis in the ALS-Penouta lab.

Table 2. Chemical composition of Bedrock and Tailings Pond head samples obtained through XRF.

Sample	Sn (ppm)	Ta (ppm)	Nb (ppm)
Penouta-Bedrock	392 ± 5	114 ± 10	31 ± 2
Penouta-Tailings Pond	334 ± 5	60 ± 10	64 ± 2

The obtained values are consistent with Polonio [30], taking into account that the tailings pond contains 4,815,307 metric tonnes of material, which, as occurs in this kind of deposits, displays a highly heterogeneous distribution of the metals of interest, in contrast to the homogeneous distribution displayed by the source rock. Furthermore, Sn, Ta and Nb values obtained for the Bedrock sample are within the range reported by [30–32].

4.2. Obtaining the Specific Rate of Breakage (S_i)

Figures 2–5 display the relationship between log ($w_i(t)/w_i(0)$) and time for 75% and 85% critical speed and ball size $d = 1.9$ cm, for Bedrock and Tailings Pond samples.

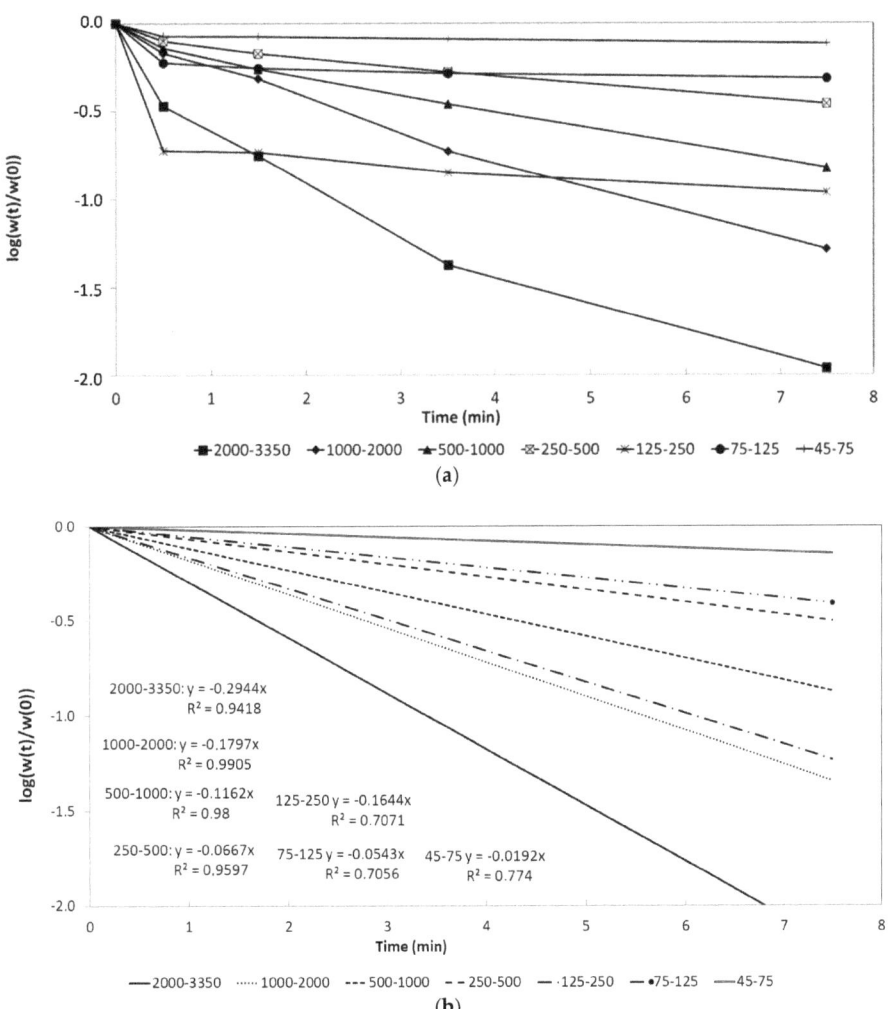

Figure 2. (a) Plot of log ($w_i(t)/w_i(0)$) vs. time for 75% critical speed and $d = 1.9$ cm (Penouta-Bedrock), (b) linear least square fitting performed.

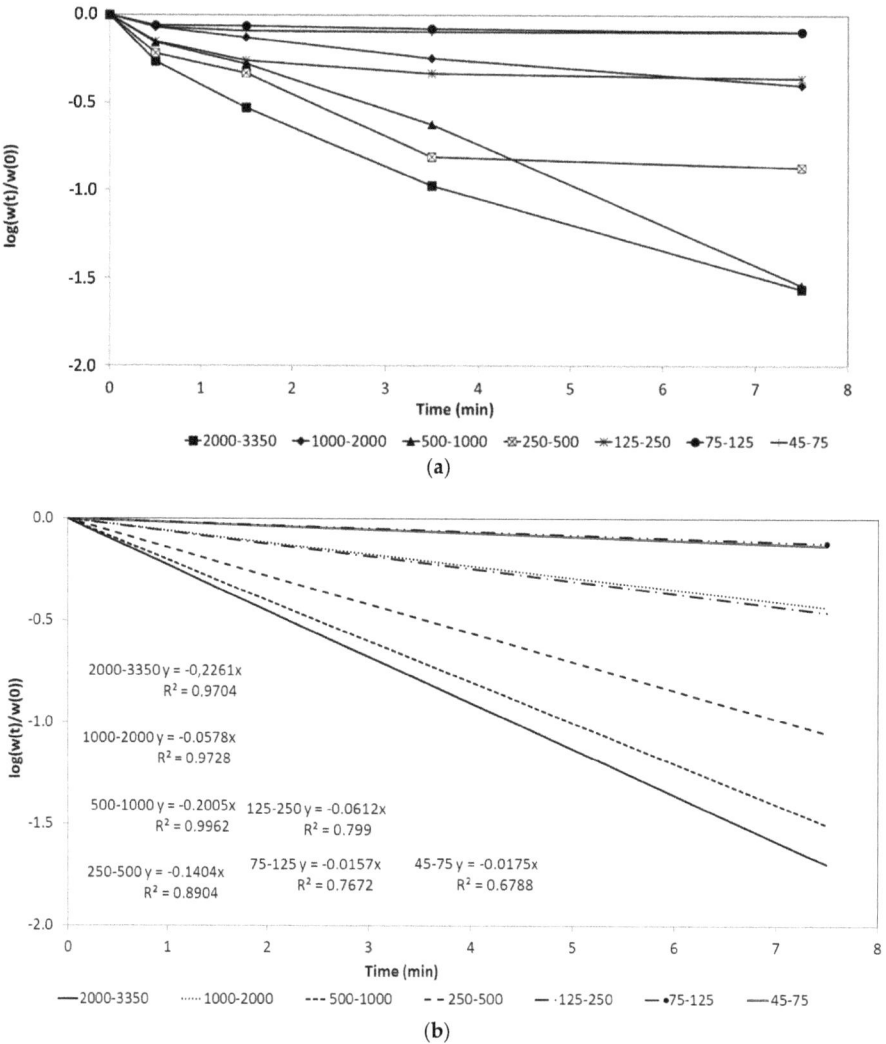

Figure 3. (a) Plot of log $(w_i(t)/w_i(0))$ vs. time; for 75% critical speed and $d = 1.9$ cm (Penouta-Tailings Pond), (b) linear least square fitting performed.

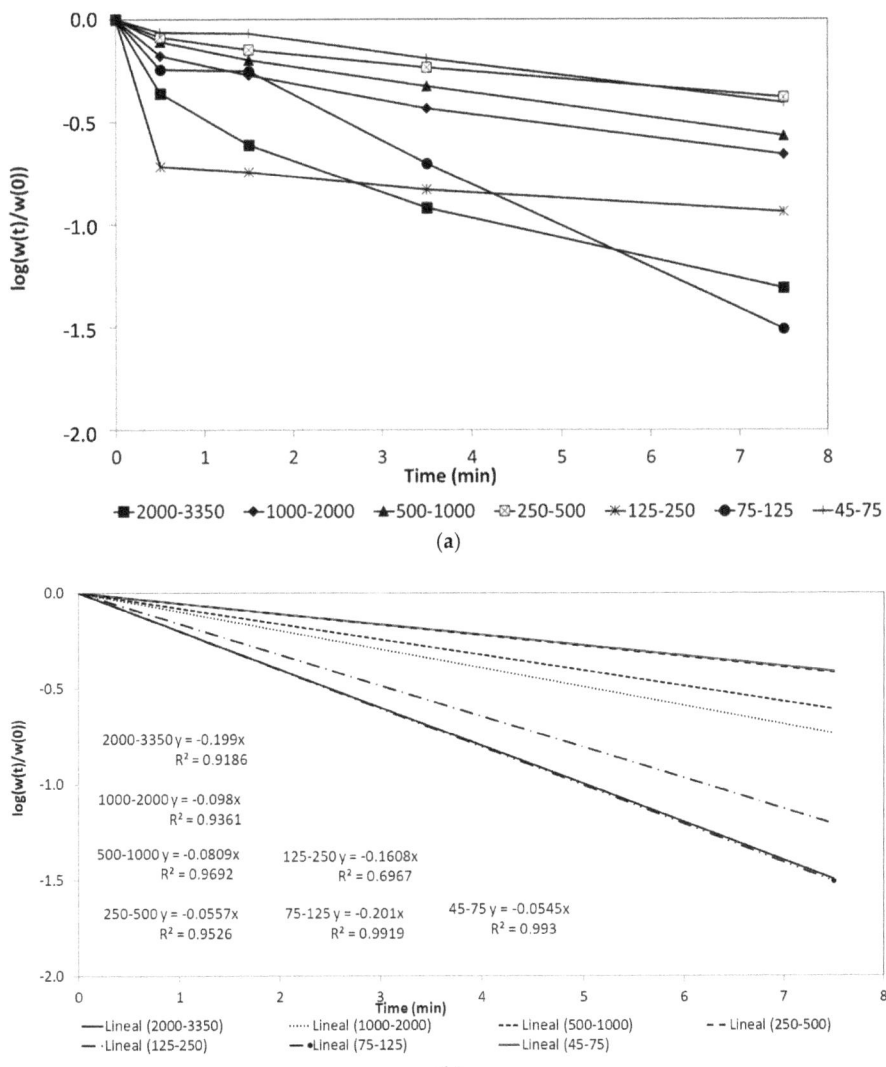

Figure 4. (**a**) Plot of log $(w_i(t)/w_i(0))$ vs. time for 85% critical speed and d = 1.9 cm (Penouta Bedrock), (**b**) linear least square fitting performed.

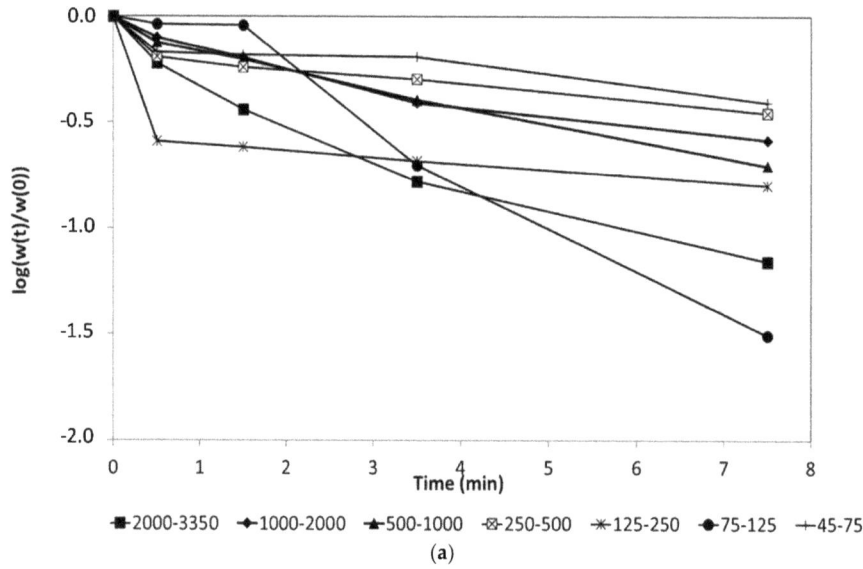

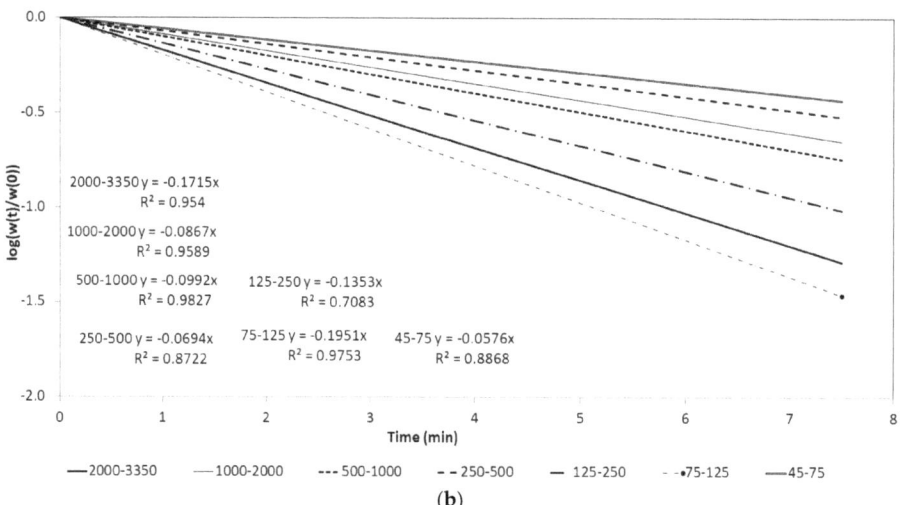

Figure 5. (a) Plot of log $(w_i(t)/w_i(0))$ vs. time for 85% critical speed and $d = 1.9$ cm (Penouta Tailings Pond), (b) linear least square fitting performed.

Figures 2–5 show a deviation from the straight lines at initial grinding stages. This is probably due to abnormal breakage and, according to [8], it should be performed a pre-grinding stage in the mill for about 2 min in order to avoid abnormal breakage behavior, which was not considered in this study.

Overall, fracture velocity of the feed monosizes fits a first order kinetic behavior, thus, being independent from time. S_i was obtained for each sample using Equation (3), and the slope calculated from Figures 2–5 for each ball-size and mill-speed condition. The relation between the specific rate of breakage S_i, and feed grain size was plotted in Figures 6–9 for each condition to visualize the behavior of S_i, as operating parameters varied for each sample.

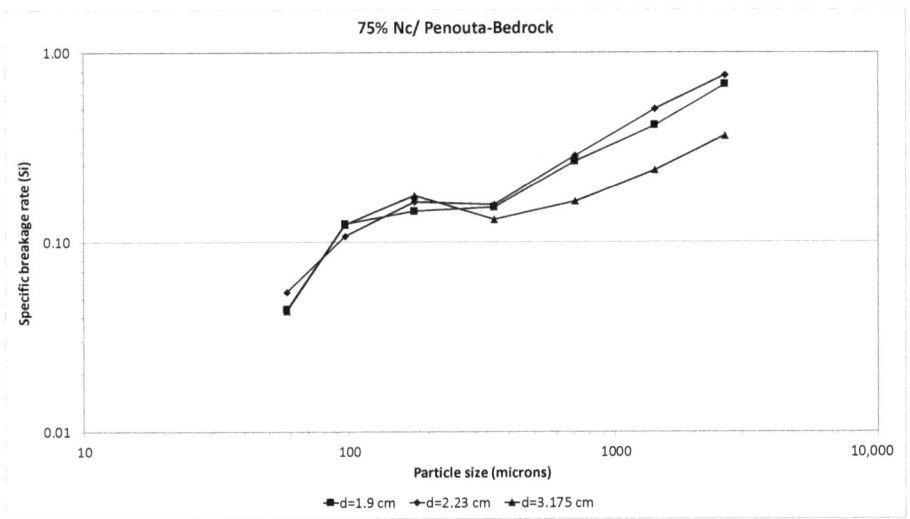

Figure 6. The specific rate of breakage vs. particle size for selected ball sizes at 75% of working speed (Penouta Bedrock).

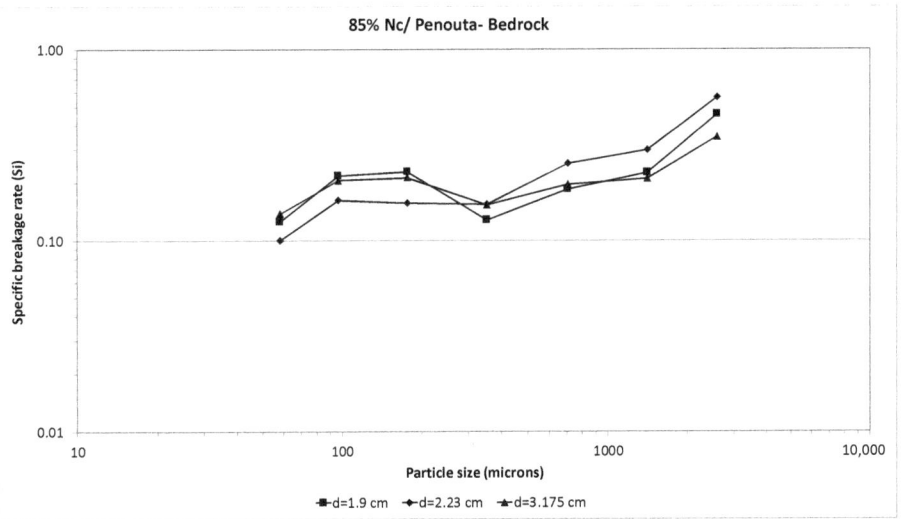

Figure 7. The specific rate of breakage vs. particle size for selected ball sizes at 85% of working speed (Penouta Bedrock).

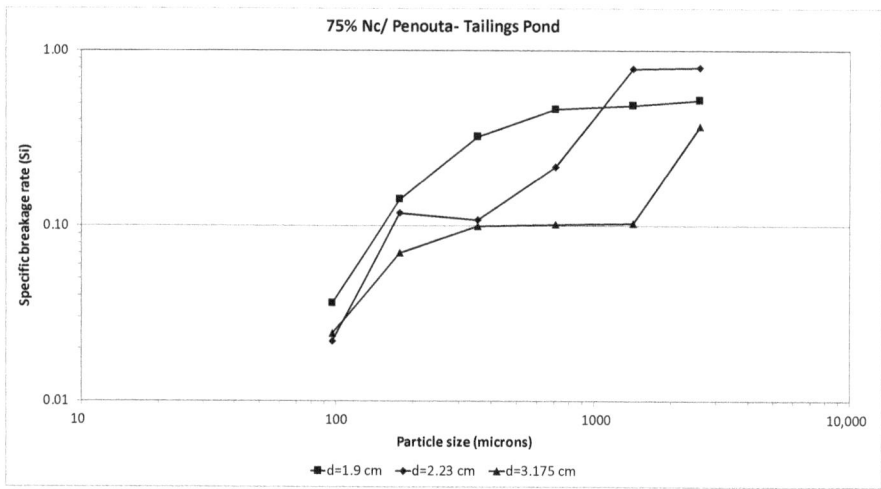

Figure 8. The specific rate of breakage vs. particle size for selected ball sizes at 75% of working speed (Penouta Tailings Pond).

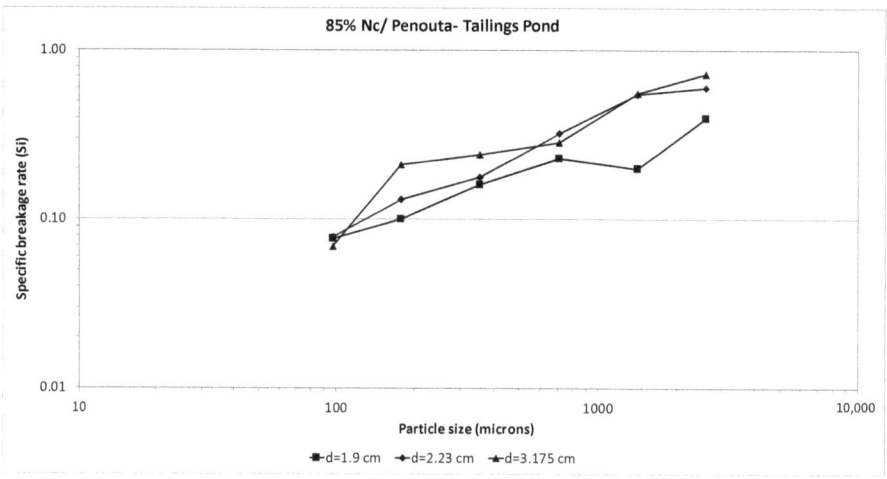

Figure 9. The specific rate of breakage vs. particle size for selected ball sizes at 85% of working speed (Penouta Tailings Pond).

In Figures 6–9, the specific rate of breakage S_i in the usual operational range increases as ball size diminishes [8,10,11,26,33–35], as happens for most of the feed grain sizes at 75% of critical mill speed. Nevertheless, at 85% critical speed, the opposite seems to happen for the Tailings Pond sample shown in Figure 9. This is probably due to better behavior under a greater influence of mill speed and ball size, mainly for the coarse feed particles size as a consequence of a greater influence of the impact breakdown and the cascading effect [36,37]. In addition, the harder ores, such as Tailings Pond samples and the coarser feeds, require high impact energy and large grinding media, and, on the other hand, very fine grind sizes require substantial grinding media surface area and small grinding media [38–40]. As a consequence, medium size balls (d = 2.23 cm) seem to have a better performance for most feed sizes, mill speeds, and samples tested [10,34,41,42].

4.3. Kinetic Parameters (α, α_T)

The grinding kinetic parameters for Bedrock and Tailings Pond samples from Penouta mine are shown in Table 3 to study the influence of ball size and mill speed in those parameters.

Table 3. Kinetic parameters for several ball sizes and mil speed (Penouta Bedrock and Tailings Pond).

Kinetic Parameters	75% Nc			85% Nc		
	d = 1.9 cm	d = 2.23 cm	d = 3.175 cm	d = 1.9 cm	d = 2.23 cm	d = 3.175 cm
α (Bedrock)	0.5531	0.6345	0.4121	0.0885	0.3764	0.1482
α_T (Bedrock)	0.0083	0.0049	0.0132	0.1451	0.0225	0.0839
α (Tailings)	0.7432	1.0414	0.6280	0.4533	0.6474	0.6320
α_T (Tailings)	0.0024	0.0003	0.0019	0.0100	0.0043	0.0053

It can be seen that α values fall within the reported normal values [26], and that the selection function α_T varies little with mill speed. From this data, the graph of Figure 10 was constructed. It plots the selection function, α_T, vs. the ball size, at constant working speed, for the studied samples.

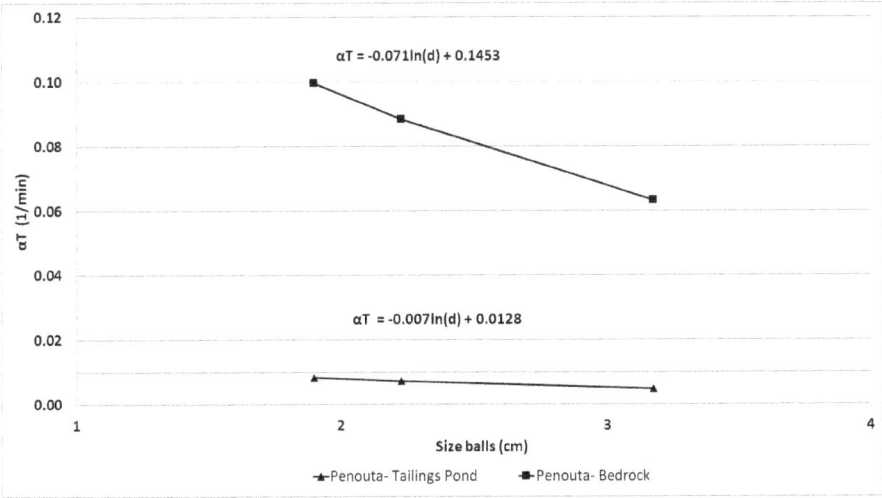

Figure 10. Graph showing the selection function vs. ball size for Penouta Bedrock and Tailings Pond.

From Table 3, the Bedrock sample yields higher α_T values than Tailings Pond sample, thus, being ground more rapidly than the latter. It must be highlighted that the Bedrock sample was taken from a slightly altered leucogranite, which results in low hardness and fracture strength. On the other hand, and due to its origin, the sandy Tailings Pond sample is heterogeneous, with a higher quartz content. It is a previously processed material and, consequently, with a higher fracture strength. In his study focused on the parameter α_T, Teke et al. [33] found a linear trend between that parameter and the ball size, characterizing the mineral calcite in this way. A good approach to determine the selection function from ball diameter in the studied samples is shown in Figure 10 with the Bedrock and Tailings Pond samples characterized through Equations (8) and (9), respectively.

$$\alpha_T = d_b + 0.1453 \tag{8}$$

$$\alpha_T = d_b + 0.0128 \tag{9}$$

where α_T is the selection function and d_b is ball size in cm.

In this sense, the results shown in Figure 10 are sound and agree with the Bond index trends previously reported for the same samples [43]. Other authors [9,44] also compared the features of other rocks like quartzite and metasandstone through the selection function, α_T.

4.4. Chemical Characterisation of the Grinding Products

The results depicted in Figures 11–14 show the relationship between the Sn yield trends and the specific rate of breakage, S_i, for each mill-speed and ball-size condition employed. Tables 4–7 include the Pearson coefficient in each case, showing a better correlation in the case of medium size balls in all cases.

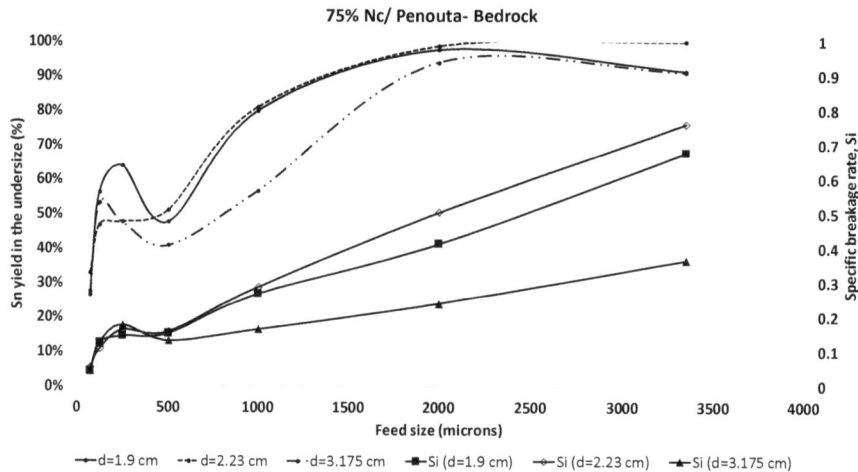

Figure 11. Plot of Sn yield and the specific rate of breakage, S_i, vs. feed size at 75% N_c for several ball sizes (Penouta Bedrock).

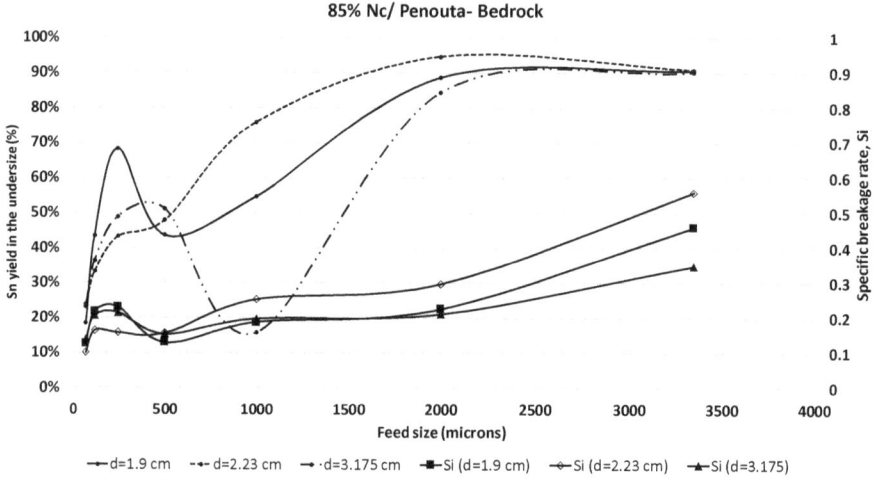

Figure 12. Plot of Sn yield and the specific rate of breakage, S_i, vs. feed size at 85% N_c for several ball sizes (Penouta Bedrock).

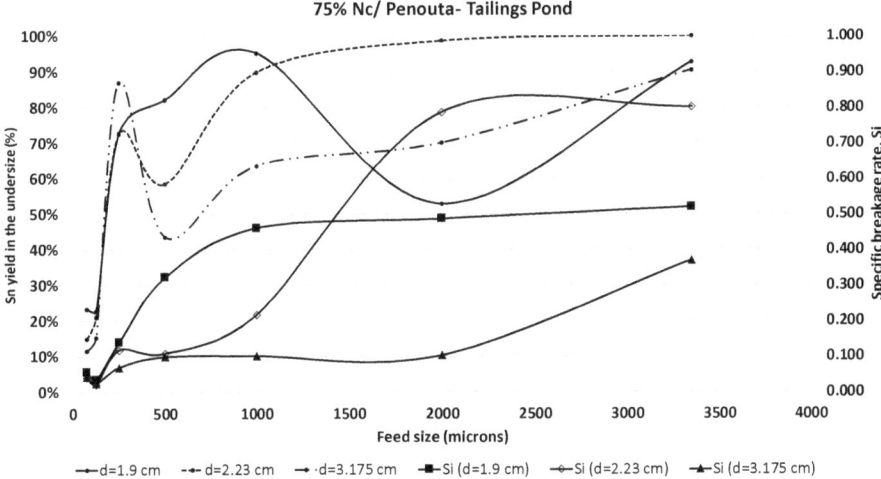

Figure 13. Plot of Sn yield and the specific rate of breakage, S_i, vs. feed size at 75% N_c for several ball sizes (Penouta Tailings Pond).

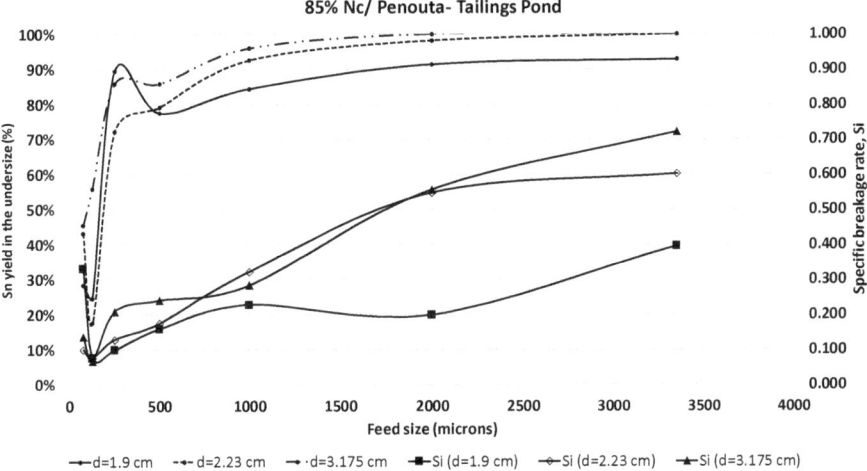

Figure 14. Plot of Sn yield and the specific rate of breakage, S_i, vs. feed size at 85% N_c for several ball sizes (Penouta Tailings Pond).

Table 4. Correlation between the specific rate of breakage S_i and Sn, Ta, Nb yields (%) for each ball size (Penouta Bedrock, 75% Nc).

Size	Ball Diameter (cm)										
	1.9				2.23				3.175		
(μm)	S_i (min^{-1})	Sn y.(%)	Ta y. (%)	Nb y. (%)	S_i (min^{-1})	Sn y. (%)	Ta y. (%)	Nb y. (%)	S_i (min^{-1})	Sn y. (%)	Ta y. (%)	Nb y. (%)
75	0.044	28%	33%	27%	0.055	33%	42%	36%	0.043	27%	29%	28%
125	0.125	56%	57%	52%	0.107	47%	46%	41%	0.124	53%	50%	50%
250	0.145	64%	45%	70%	0.163	48%	55%	49%	0.176	48%	56%	51%
500	0.153	48%	76%	63%	0.158	51%	75%	78%	0.131	41%	66%	66%
1000	0.267	80%	89%	84%	0.287	81%	88%	85%	0.164	57%	72%	66%
2000	0.413	98%	99%	93%	0.504	99%	97%	93%	0.24	94%	93%	84%
3350	0.677	91%	68%	92%	0.761	100%	100%	100%	0.365	91%	71%	90%
Pearson c. (r)		0.83	0.51	0.80		0.92	0.87	0.84		0.89	0.68	0.89

Table 5. Correlation between the specific rate of breakage S_i and Sn, Ta, Nb yields (%) for each ball size (Penouta Bedrock, 85% Nc).

Size	Ball Diameter (cm)										
	1.9				2.23				3.175		
(μm)	S_i (min^{-1})	Sn y. (%)	Ta y. (%)	Nb y. (%)	S_i (min^{-1})	Sn y. (%)	Ta y. (%)	Nb y. (%)	S_i (min^{-1})	Sn y. (%)	Ta y. (%)	Nb y. (%)
75	0.125	18%	19%	18%	0.1	23%	23%	23%	0.138	24%	30%	23%
125	0.217	43%	45%	45%	0.163	33%	37%	34%	0.206	36%	41%	38%
250	0.229	68%	70%	71%	0.157	43%	50%	46%	0.214	49%	52%	52%
500	0.128	43%	65%	62%	0.155	48%	68%	58%	0.153	51%	72%	66%
1000	0.186	55%	75%	67%	0.252	76%	87%	81%	0.197	16%	29%	23%
2000	0.225	89%	89%	79%	0.297	95%	93%	86%	0.211	84%	83%	75%
3350	0.458	91%	74%	93%	0.559	91%	70%	91%	0.348	90%	61%	88%
Pearson c. (r)		0.75	0.41	0.70		0.81	0.52	081		0.68	0.25	0.64

Table 6. Correlation between the specific rate of breakage S_i and Sn, Ta, Nb yields (%) for each ball size (Penouta Tailings Pond, 75% Nc).

Size	Ball Diameter (cm)										
	1.9				2.23				3.175		
(μm)	S_i (min^{-1})	Sn y. (%)	Ta y. (%)	Nb y. (%)	S_i (min^{-1})	Sn y. (%)	Ta y. (%)	Nb y. (%)	S_i (min^{-1})	Sn y. (%)	Ta y. (%)	Nb y. (%)
75	0.057	23%	35%	28%	0.041	15%	21%	16%	0.043	11%	17%	13%
125	0.036	23%	33%	26%	0.022	21%	31%	21%	0.024	15%	15%	15%
250	0.141	73%	60%	60%	0.118	73%	60%	56%	0.070	87%	83%	82%
500	0.323	82%	96%	90%	0.108	58%	76%	71%	0.099	44%	70%	58%
1000	0.461	95%	81%	90%	0.216	90%	78%	84%	0.101	64%	63%	64%
2000	0.487	53%	70%	67%	0.786	99%	93%	97%	0.103	70%	58%	44%
3350	0.520	93%	70%	79%	0.800	100%	100%	100%	0.370	90%	89%	89%
Pearson c. (r)		0.75	0.75	0.83		0.78	0.80	0.80		0.64	0.65	0.68

Table 7. Correlation between the specific rate of breakage S_i and Sn, Ta, Nb yields (%) for each ball size (Penouta Tailings Pond, 85% Nc).

Size	Ball Diameter (cm)											
	1.9				2.23				3.175			
(µm)	S_i (min^{-1})	Sn y. (%)	Ta y. (%)	Nb y. (%)	S_i (min^{-1})	Sn y. (%)	Ta y. (%)	Nb y. (%)	S_i (min^{-1})	Sn y. (%)	Ta y. (%)	Nb y. (%)
75	0.331	28%	31%	30%	0.100	43%	58%	48%	0.138	46%	64%	51%
125	0.077	25%	27%	23%	0.078	18%	27%	20%	0.069	56%	71%	62%
250	0.100	90%	85%	83%	0.130	72%	61%	56%	0.210	86%	82%	78%
500	0.160	78%	82%	75%	0.176	79%	90%	67%	0.241	86%	98%	90%
1000	0.228	84%	81%	84%	0.323	93%	92%	90%	0.284	96%	92%	90%
2000	0.199	91%	75%	75%	0.548	98%	95%	91%	0.557	100%	100%	100%
3350	0.395	93%	89%	89%	0.601	100%	100%	100%	0.721	100%	100%	100%
Pearson c. (r)		0.13	0.13	0.19		0.80	0.79	0.88		0.76	0.77	0.81

Finally, Figures 15 and 16 depict the plot of Sn, Ta and Nb yield in the undersize product, vs. the specific rate of breakage S_i, at 75% mill critical speed and ball size of 2.23 cm for both studied samples.

The results depicted in Figures 11–16 demonstrate that direct relationships exist amongst Sn, Ta and Nb yield in the undersize product, as elements of interest in the product, the specific rate of breakage and the operational variables mill speed, ball size and feed size. Consequently, it can be stated that, at 75% of critical speed, grinding is more efficient with medium to small ball sizes, whereas, at 85% of critical speed, better results occur with larger ball sizes. These conditions would represent the optimal working parameters to enhance the specific rate of breakage, thus, guaranteeing a proper mineral liberation and concomitantly a higher mineral recovery and product grade.

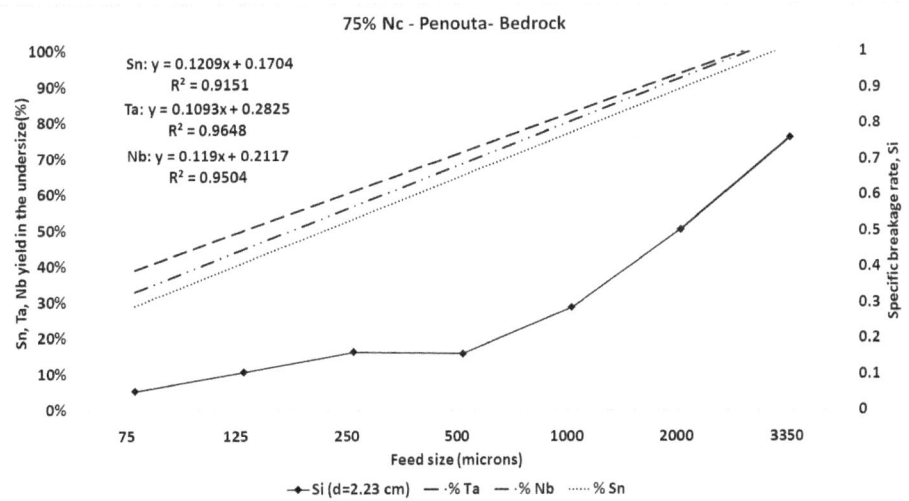

Figure 15. Plot of Sn, Ta and Nb yield in the undersize product and the specific rate of breakage, S_i, vs. feed size for 75% N_c and ball size = 2.23 cm (Penouta Bedrock).

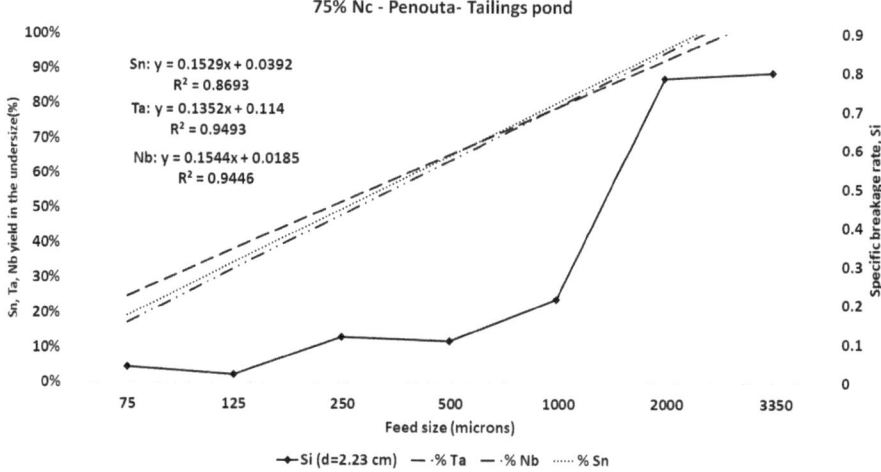

Figure 16. Plot of Sn, Ta and Nb yield in the undersize product and the specific rate of breakage, S_i, vs. feed size for 75% N_c and ball size = 2.23 cm (Penouta Tailings Pond).

5. Conclusions

The experimental work done and its further analysis permit to draw the following conclusions:

- Austin's methodology has allowed studying the effects of ball size in the kinetics of dry and batch grinding over a wide range of feed particle size feed for the samples Bedrock and Tailings Pond (Penouta mine). The mineralogical and operational parameters studied in this investigation, mill speed, ball size and feed size, also influenced the grinding kinetics.
- S_i decreases as feed particle size decreases and ball size increases. This is due to a reduction of the effective grinding area over most conditions considered, and to the fact that the finer the particle size the higher the fracture strength, owing to the lesser crack and microcrack concentration in the particles.
- A direct relation exists amongst Sn, Ta and Nb yield in the undersize product, the S_i and the studied mineralogical and operational variables. Optimal mineralogical and operational conditions will increase the grinding efficiency to obtain the best liberation degree and the highest grade of minerals of interest, such as Sn, Ta and Nb, thus impacting positively the recovery scores of the plant.
- Use of medium-diameter balls is recommended, since they yield a steadier behavior over a wide range of feed particle sizes and studied conditions.
- Using ball size, a selection function, α_T, was formulated for the Bedrock and Tailings Pond samples from the Penouta mine. This demonstrated that α_T values are higher for Bedrock sample than for Tailings Pond sample, resulting in the former being ground more rapidly than the latter, as a consequence of their respective mineralogy and origin.

Author Contributions: Conceptualization and execution of experiments, J.V.N.; methodology, J.V.N.; formal analysis, J.V.N., T.L., J.M.M.-A.; investigation, J.V.N.; data curation, J.V.N.; writing–original draft preparation, J.V.N.; writing–review and editing, J.V.N., T.L. and J.M.M.-A.; supervision, T.L. and J.M.M.-A.; and project administration and funding acquisition, J.M.M.-A. All authors have read and agreed to the published version of the manuscript.

Funding: This work is part of the OptimOre project funded by the European Union Horizon 2020 Research and Innovation Programme under grant agreement No 642201.

Acknowledgments: The authors thank Strategic Minerals Spain, S.L. for their support providing the samples.

Conflicts of Interest: The authors declare no conflict of interest.

References

1. Fuerstenau, D.W.; Lutch, J.J.; De, A. The effect of ball size on the energy efficiency of hybrid high-pressure roll mill/ball mill grinding. *Powder Technol.* **1999**, *105*, 199–204. [CrossRef]
2. Coello Velázquez, A.L.; Menéndez-Aguado, J.M.; Brown, R.L. Grindability of lateritic nickel ores in Cuba. *Powder Technol.* **2008**, *182*, 113–115. [CrossRef]
3. Pedrayes, F.; Norniella, J.G.; Melero, M.G.; Menéndez-Aguado, J.M.; del Coz-Díaz, J.J. Frequency domain characterization of torque in tumbling ball mills using DEM modelling: Application to filling level monitoring. *Powder Technol.* **2018**, *323*, 433–444. [CrossRef]
4. Osorio, A.M.; Menéndez-Aguado, J.M.; Bustamante, O.; Restrepo, G.M. Fine grinding size distribution analysis using the Swebrec function. *Powder Technol.* **2014**, *258*, 206–208. [CrossRef]
5. Rodríguez, B.Á.; García, G.G.; Coello-Velázquez, A.L.; Menéndez-Aguado, J.M. Product size distribution function influence on interpolation calculations in the Bond ball mill grindability test. *Int. J. Miner. Process.* **2016**, *157*, 16–20. [CrossRef]
6. Deniz, V. The effect of mill speed on kinetic breakage parameters of clinker and limestone. *Cem. Concr. Res.* **2004**, *34*, 1365–1371. [CrossRef]
7. Olejnik, T.P. Analysis of the breakage rate function for selected process parameters in quartzite milling. *Chem. Process. Eng. Inzynieria Chemiczna i Procesowa* **2012**, *33*, 117–129. [CrossRef]
8. Gupta, V.K.; Sharma, S. Analysis of ball mill grinding operation using mill power specific kinetic parameters. *Adv. Powder Technol.* **2014**, *25*, 625–634. [CrossRef]
9. Petrakis, E.; Komnitsas, K. Improved modeling of the grinding process through the combined use of matrix and population balance models. *Minerals* **2017**, *7*, 67. [CrossRef]
10. Cayirli, S. Influences of operating parameters on dry ball mill performance. *Physicochem. Probl. Mineral Process.* **2018**, *54*, 751–762.
11. Cho, H.; Kwon, J.; Kim, K.; Mun, M. Optimum choice of the make-up ball sizes for maximum throughput in tumbling ball mills. *Powder Technol.* **2013**, *246*, 625–634. [CrossRef]
12. Hlabangana, N.; Danha, G.; Muzenda, E. Effect of ball and feed particle size distribution on the milling efficiency of a ball mill: An attainable region approach. *S. Afr. J. Chem. Eng.* **2018**, *25*, 79–84. [CrossRef]
13. Bwalya, M.M.; Moys, M.H.; Finnie, G.J.; Mulenga, F.K. Exploring ball size distribution in coal grinding mills. *Powder Technol.* **2014**, *257*, 68–73. [CrossRef]
14. King, R.P. *Modeling and Simulation of Mineral Processing Systems*; Butterworth-Heinemann Elsevier Ltd.: Oxford, UK, 2001.
15. Epstein, B. The mathematical description of certain breakage mechanisms leading to the logarithmic-normal distribution. *J. Frankl. Inst.* **1947**, *244*, 471–477. [CrossRef]
16. Kelsall, D.F. A study of the breakage in a small continuous open circuit wet mill. *Can. Min. J.* **1965**, *86*, 89–94.
17. Herbst, J.A.; Fuerstenau, D.W. The zero-order production of fine sizes in comminution and its implications in simulation. *Trans. AIME* **1968**, *241*, 538–549.
18. Austin, L.G.; Luckie, P.T. Methods for the determination of breakage distribution parameters. *Powder Technol.* **1971**, *5*, 267–271. [CrossRef]
19. Austin, L. A Review–Introduction to the mathematical description of grinding as a rate process. *Powder Technol.* **1972**, *5*, 1–17. [CrossRef]
20. Gardner, R.P.; Sukanjnajtee, K. A combined tracer and back-calculation method for determining particulate breakage functions in ball milling: Part I—rationale and description of the proposed method. *Powder Technol.* **1972**, *6*, 65–74. [CrossRef]
21. Kapur, P.C. An improved method for estimating the feed-size breakage distribution functions. *Powder Technol.* **1982**, *33*, 269–275. [CrossRef]
22. Kapur, P.C.; Agrawal, P. Approximate solutions to the discretized batch grinding equation. *Chem. Eng. Sci.* **1970**, *25*, 1111–1113. [CrossRef]
23. Malghan, S.G.; Fuerstenau, D.W. The scale-up of ball mills using population balance models and specific power input. *Dechem. Monogr.* **1976**, *79*, 613–630.

24. Austin, L.G.; Luckie, P.T.; Shoji, K. An analysis of ball-and-race milling part II: The babcock E 1.7 mill. *Powder Technol.* **1982**, *33*, 113–125. [CrossRef]
25. Austin, L.G.; Brame, K. A comparison of the Bond method for sizing wet tumbling ball mills with a size—mass balance simulation model. *Powder Technol.* **1983**, *34*, 261–274. [CrossRef]
26. Austin, L.; Concha, F. *Diseño y Simulación de Circuitos de Molienda y Clasificación*; CYTED: Valparaiso, Chile, 1994.
27. Austin, L.G.; Klimpel, R.R.; Luckie, P. *Process Engineering of Size Reduction: Ball Milling*; Society of Mining Engineers of the AIME: Littleton, CO, USA, 1984.
28. European Commission. Study on the EU's List of Critical Raw Materials. Available online: https://ec.europa.eu/commission/presscorner/detail/en/ip_20_1542 (accessed on 1 January 2020).
29. Wang, X.; Gui, W.; Yang, C.; Wang, Y. Wet grindability of an industrial ore and its breakage parameters estimation using population balances. *Int. J. Miner. Process.* **2011**, *98*, 113–117. [CrossRef]
30. Polonio, G.F. El Interés Económico y Estratégico del Aprovechamiento de Metales Raros y Minerales Industriales Asociados, en el Marco Actual de la Minería Sostenible: La Mina de Penouta. Ph.D. Thesis, Universidad Politécnica de Madrid, Orense, España, 2015.
31. Llorens, G.T.; García, P.F.; López, M.F.J.; Fernández, F.A.; Sanz, C.J.L.; Moro, B.M.C. Tin-tantalum-niobium mineralization in the Penouta deposit (NW Spain): Textural features and mineral chemistry to unravel the genesis and evolution of cassiterite and columbite group minerals in a peraluminous system. *Ore Geol. Rev.* **2017**, *81*, 79–95. [CrossRef]
32. López-Moro, F.J.; García, P.F.; Llorens, G.T.; Sanz, C.J.L.; Fernández, F.A.; Moro, B.M.C. Ta and Sn concentration by muscovite fractionation and degassing in a lens-like granite body: The case study of the Penouta rare-metal albite granite (NW Spain). *Ore Geol. Rev.* **2017**, *82*, 10–30. [CrossRef]
33. Teke, E.; Yekeler, M.; Ulusoy, U.; Canbazoglu, M. Kinetics of dry grinding of industrial minerals: Calcite and barite. *Int. Miner. Process.* **2002**, *67*, 29–42. [CrossRef]
34. Erdem, A.S.; Ergün, S.L. The effect of ball size on breakage rate parameter in a pilot scale ball mill. *Miner. Eng.* **2009**, *22*, 660–664. [CrossRef]
35. Gupta, V.K. Determination of the specific breakage rate parameters using the top-size-fraction method: Preparation of the feed charge and design of experiments. *Adv. Powder Technol.* **2016**, *27*, 1710–1718. [CrossRef]
36. Touil, D.; Belaadi, S.; Frances, C. The specific selection function effect on clinker grinding efficiency in a dry batch ball mill. *Int. J. Miner. Process.* **2008**, *87*, 141–145. [CrossRef]
37. De Carvalho, R.M.; Tavares, L.M. Predicting the effect of operating and design variables on breakage rates using the mechanistic ball mill model. *Miner. Eng.* **2013**, *43–44*, 91–101. [CrossRef]
38. Napier-Munn, T.J.; Morrell, S.; Morrison, R.D.; Kojovic, T. *Mineral. Comminution Circuits: Their Operation and Optimization*; Napier-Munn, T.J., Ed.; Julius Kruttschnitt Mineral Research Centre, University of Queensland: Indooroopilly, QLD, Australia, 1996.
39. Wills, B.A.; Napier-Munn, T. Mineral Processing Technology. In *An Introduction to the Practical Aspects of Ore Treatment and Mineral Recovery*, 7th ed.; Elsevier Science & Technology Books: London, UK, 2006.
40. Chimwani, N.; Glasser, D.; Hildebrandt, D.; Metzger, M.J.; Mulenga, F.K. Determination of the milling parameters of a platinum group minerals ore to optimize product size distribution for flotation purposes. *Miner. Eng.* **2013**, *43–44*, 67–78. [CrossRef]
41. Mulenga, F. Effect of Ball Size Distribution on Milling Parameters. Ph.D. Thesis, University of Witwatersrand, Johannesburg, South Africa, 2008.
42. Ucurum, M.; Gulec, Ö.; Cingitas, M. Wet grindability of calcite to ultra-fine sizes in conventional ball mill. *Part. Sci. Technol.* **2015**, *33*, 342–348. [CrossRef]
43. González, G.; Menéndez, A.J.M. Variación del índice de trabajo en molino de bolas según el grado de molienda para varias menas de tungsteno. In Proceedings of the XIII Jornadas Argentinas de Tratamiento de Minerales Octubre de 2016, Mendoza, Argentina, 5–7 October 2016.

44. Petrakis, E.; Stamboliadis, E.; Komnitsas, K. Identification of Optimal Mill Operating Parameters during Grinding of Quartz with the Use of Population Balance Modeling. *KONA Powder Part. J.* **2017**, *34*, 213–223. [CrossRef]

Publisher's Note: MDPI stays neutral with regard to jurisdictional claims in published maps and institutional affiliations.

© 2020 by the authors. Licensee MDPI, Basel, Switzerland. This article is an open access article distributed under the terms and conditions of the Creative Commons Attribution (CC BY) license (http://creativecommons.org/licenses/by/4.0/).

Article
Grinding Kinetics Study of Tungsten Ore

Jennire V. Nava [1], Alfredo L. Coello-Velázquez [2] and Juan M. Menéndez-Aguado [1,*]

1. Escuela Politécnica de Mieres, Universidad de Oviedo, C/Gonzalo Gutiérrez Quirós, 33600 Mieres, Asturias, Spain; jvanessanavar@gmail.com
2. CETAM, Universidad de Moa Dr. Antonio Núñez Jiménez, Moa 83300, Cuba; acoello@ismm.edu.cu
* Correspondence: maguado@uniovi.es

Abstract: The European Commission (EC) maintains the consideration of tungsten as a critical raw material for the European industry, being the comminution stage of tungsten-bearing minerals an essential step in the tungsten concentration process. Comminution operations involve approximately 3–4% of worldwide energy consumption; therefore, grinding optimization should be a priority. In this study, the grinding behavior of tungsten ore from Barruecopardo Mine (Salamanca, Spain) is analyzed. A protocol based on Austin's methodology and PBM is developed in order to study the influence of operational and geometallurgical variables on grinding kinetics. In addition to the kinetic parameters, the breakage probability (S_i) and breakage function (B_{ij}) is determined. The selection function was formulated for the Barruecopardo Mine with respect to the mill speed.

Keywords: critical raw materials; tungsten ore; grinding kinetics

1. Introduction

The European Union has recently published the updated list of critical raw materials, in which tungsten (W) is included. This critical condition is defined by both the supply risk to the EU and the economic importance developed on the industrial value chains of the European Union [1].

Tungsten presents strategic applications on high strength alloys for machining tools, automotive and mobile phone sectors, among others [2]. Currently, there are several tungsten mines in Europe, some active and others on the exploration stage [3]. This is the case of the Barruecopardo mine in Salamanca (Spain), owned by Ormonde Mining PLC and currently administered by Saloro S.L., which is estimated to provide 11% of the non-Chinese global supply of tungsten [4]. The main minerals present in Barruecopardo are scheelite ($CaWO_4$) and wolframite (($Fe, Mn)WO_4$), which constitute the ore. Arsenopyrite (FeAsS), pyrite (FeS_2), chalcopyrite ($CuFeS_2$), and ilmenite ($FeTiO_3$) are also present as primary minerals of the gangue [5,6].

Regarding mineral benefit, the comminution stage represents 3–4% of the energy consumption worldwide and 40–70% of the energy consumed in a mineral processing plant [7,8]. In fact, ball-mill grinding is one of the most energy-consuming techniques. Therefore, setting the optimal values of the operational and mineralogical parameters both for the initial design and the process adaptation to ore variations [9].

Several researchers have investigated the influence exerted on kinetic conditions by operational parameters such as mill speed [10,11] and filling volume [12,13]. Other researchers devoted their work to study geometallurgical variables such as grain size, shape and roughness, specific surface area, orientation, hardness, fracture strength, feed particle size distribution, and mineralogy [14–20] using optical microscopy or more advanced techniques, such as Quantitative Microstructural Analysis (QMA) [21]. Consequently, a small improvement in machinery efficiency and an optimal design in the grinding system, taking into account the optimization of the above-mentioned parameters, would greatly cut down plant operational costs, impacting environmental issues and resource management [22,23].

This work aims to characterize the grinding kinetic behavior at a lab-scale of tungsten ore, due to its importance as a critical raw material to the EU, by determining the kinetic parameters following the Austin model. A second objective is to assess the influence of mill speed on the kinetic and geometallurgical parameters, being mill speed the more easily adjustable operational parameter at an industrial scale.

2. Theoretical Background

The population balance model (PBM) has been widely used in ball mills since its proposal by Austin [24]. This model is based on determining the particle size distribution grouped in size classes. A mass balance for the class i in a well-mixed grinding process is done by employing Equation (1), where a first-order kinetic fragmentation is assumed.

$$\frac{dw_i}{dt} = -S_i w_i(t) + \sum_{j=1}^{i-1} b_{ij} k_j w_j(t) \tag{1}$$

where $w_i(t)$ is the remnant mass fraction of particle size class i at grinding time t. The first term of the right-hand side is the mass fraction of the monosize i particles that break and thus no longer belong to that monosize, being S_i the probability of fracture. The second term represents the contribution of all monosizes coarser than i that break to produce particles of monosize i. The fracture rate of a monosize material can be expressed by Equation (2):

$$\frac{-dw_i}{dt} = S_i w_i(t) \tag{2}$$

where S_i is the probability of fracture or specific fracture rate, whose unit is t^{-1}. Assuming that S_i does not change with time, the integral results in Equation (3).

$$Log(wi(t)) - Log(wi(0)) = \frac{-S_i(t)}{2.3} \tag{3}$$

where w_i is the weight fraction of mineral feed into the mill having a size 1 for time t, and S_i is the probability of fracture. According to the methodology proposed by Austin et al. [25], once S_i values have been obtained through slope determination, they are plotted to the particle size, and Equation (4) is proposed to study the behavior of the probability of fracture S_i.

$$S_i = \alpha_T X_i^\alpha \tag{4}$$

where X_i is the upper size limit of the interval in mm, and α_T and α are model parameters that depend on the material properties and the grinding conditions. To find the second term of Equation (1), the fracture function b_{ij} is defined. This function represents the particle fraction that belongs initially to interval j, after fracture falls in interval i. It is recommended to represent this value in cumulative form B_{ij}, whose calculation is done with Equation (5).

$$B_{ij} = \sum_{k=n}^{i} b_{kj} \tag{5}$$

That is, B_{ij} is the sum of the mineral fractions finer than the upper limit of interval i as a result of the primary break of the size interval j. Austin et al. [25] showed that B_{ij} values could be estimated from a size analysis of the products over short grinding times of an initial feed chiefly of size j through the method BII [25–27]. With the parameters of fracture function, B_{ij} can be determined graphically through an empirical function like Equation (6).

$$B_{ij} = \phi_j \left(\frac{X_{i-1}}{X_j}\right)^\gamma + (1 - \phi_j)\left(\frac{X_{i-1}}{X_j}\right)^\beta \quad n \geq i \geq j+1 \tag{6}$$

where ϕ_j, γ and β are parameters that depend on the material properties. The critical speed, Nc, is calculated using Equation (7).

$$Nc = \frac{42.3}{\sqrt{D-d}} \qquad (7)$$

where D is the mill diameter and d is the ball diameter [m]. Ball mill filling volume is calculated using Equation (8).

$$J = \left(\frac{mass\ of\ balls}{ball\ density \times mill\ volume}\right) \times \frac{1.0}{0.6} \qquad (8)$$

On the other side, Austin and Brame [28] calculated the sorting function α_T in a general way through Equation (9).

$$\alpha_T = \frac{v_c - 0.1}{1 + e^{[15.7(v_c - 0.94)]}} \qquad (9)$$

where v_c is the mill speed expressed as a fraction of the critical speed. Finally, according to [29], the Froude number expresses the ratio of centrifugal acceleration to gravity acceleration at the perimeter of the mill chamber (Equation (10)). This number can be used to characterize the charge motion in the mill and the ball regime. Thus, in laboratory ball mills, it is recommended to define work conditions with centrifugal acceleration at the shell equalling 1/2 of the acceleration due to gravity ($F_r = 0.5$), corresponding to $v_c = 70.7\%$ [29].

$$F_r = \frac{\frac{D}{2}\omega^2}{g} = \frac{2\pi^2 n^2 D}{g} \qquad (10)$$

3. Materials and Methods

3.1. Sample Preparation and Feed Characterization

The sample mineralogy was extensively characterized by Alfonso et al. [6]. A representative sample from an old waste dump of the Barruecopardo mine was prepared in a 4 mm jaw crusher. After homogenization and quartering using a riffle splitter, sieving provided an adequate quantity of the following size intervals, which will be considered as monosizes in this work: 5000/4000, 4000/3350, 3000/2000, 2000/1000, 1000/500, 500/250, 250/125, 125/75, 75/45 µm.

A representative sample was characterized chemically through XRF, using a Bruker XRF S-4 Pioneer Advance, with sample preparation in a Claisse Perler, model M-4.

3.2. Calculation of Critical Speed and Initial Conditions for the Grinding Kinetic Tests

Mill critical speed was calculated using Equation (7). Table 1 shows the three milling speeds used in the tests.

Table 1. Mill rotation speeds used in the grinding kinetic tests.

Mill Speed	n [rpm]	Froude Number
Nc ($v_c = 1$)	112.3	1
N_1 ($v_c = 0.6$)	67.4	0.43
N_2 ($v_c = 0.7$)	78.6	0.58
N_3 ($v_c = 0.8$)	89.9	0.76

Grinding kinetic tests were run in a laboratory mill, 17.8 cm in diameter and 4500 cm^3 in capacity. Feed was of 900 cm^3, and milling load consisted of 6.6 kg of steel balls with the following ball size distribution: 45 balls 19 mm in diameter, 23 balls 29.7 mm in diameter, and 17 balls 36.8 mm in diameter. Fill fraction was calculated using Equation (8). The feed consisted of samples of the 9 monosizes selected (5000/4000, 4000/3000, 3000/2000, 2000/1000, 1000/500, 500/250, 250/125, 125/75, 75/45 µm). Grinding times were 0.5, 1,

1.5, 3.5, 6.5, and 10.5 min. Each sample was dumped and, after performing a grain size analysis, W content was measured to assess the evolution of the W grade with respect to the kinetic parameters.

3.3. Determination of Fracture Probability (S_i), Fracture Function (B_{ij}), and the Kinetic Parameters (α_T, α, ϕ_j, γ, and β)

3.3.1. Fracture Probability (S_i) and Kinetic Parameters (α_T, and α)

After obtaining the oversize weights for each grinding time, and plotting the time function log ($w_i(t)/w_i(0)$) for each monosize, the equation for each curve, and consequently, the S_i value, were calculated through the linear fitting using Equation (3). Then, S_i, values for each monosize were plotted, and the parameters (α_T and α) were calculated using Equation (4) for each mill speed condition to study the influence of this operational variable on the probability of fracture and on the kinetic parameters α_T and α. The selection function α_T, obtained through Equation (9) was calculated using an equation designed for this particular ore, as detailed in Section 4.2.

3.3.2. Determination of the Fracture Function (B_{ij}) and the Kinetic Parameters (ϕ_j, γ and β)

These calculations were done for each feed monosize, i, at each mill speed condition and after the grinding time. Product particle size analysis was done by sieving with mesh sizes i to $j-n$. Weight of oversizes i and S_{j-n} was determined, and fracture function values b_{ij} and cumulative value B_{ij}, were obtained using Equation (5). Finally, B_{ij} data were plotted to relative size j/i for each monosize and mill speed conditions. The rest of the kinetic parameters were calculated through Equation (6).

3.4. Study of P_{80} and the Ratio of Reduction R_r

The evolution of some relevant parameters of the product was represented. These parameters were P_{80} and the ratio of reduction, R_r, after a 0.5 min grinding time at several mill speeds for each feed monosize. Likewise, the evolution of the probability of fracture, S_i, and R_r with mill speed was studied, for each feed monosize.

3.5. Chemical Characterisation of the Product

Milling products were chemically characterized with the equipment specified in Section 3.1.

4. Results and Discussion

4.1. Feed Characterzsation

The sample mechanically prepared displays a particle size distribution (PSD), as shown in Figure 1.

F_{80} for the Barruecopardo sample is 1690 μm. Table 2 shows W contents obtained from a mineralogical study for each size interval.

Table 2. Fractional results of XRF (ND = not detected).

Size Interval (μm)	Weight (%)	W (ppm)	STD (ppm)
>4000	1.79	ND	-
3350–4000	3.93	ND	-
2000–3350	9.71	ND	-
1000–2000	29.53	40	10
500–1000	24.01	105	13
250–500	19.94	146	13
125–250	8.23	75	12
75–125	1.52	23	10
45–75	0.78	ND	-
<45	0.55	ND	-

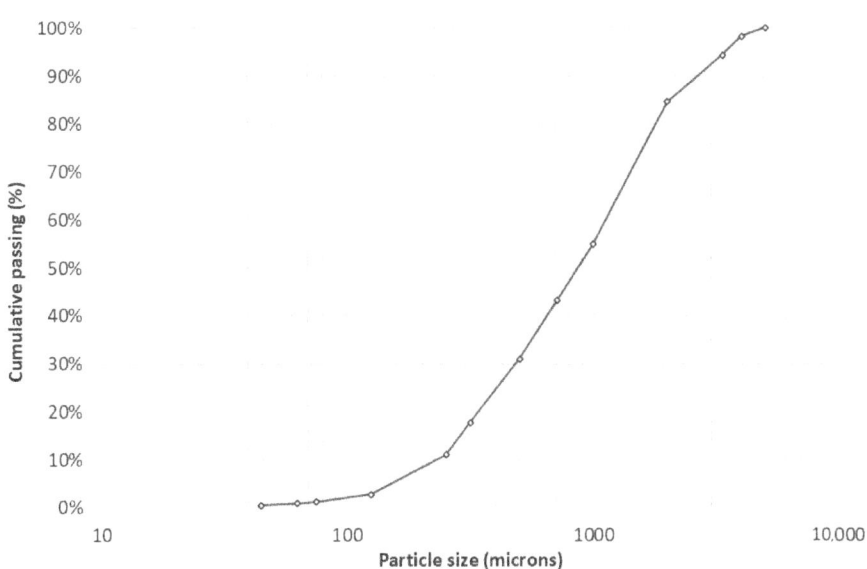

Figure 1. PSD for the Barruecopardo ore head samples.

W contents shown in Table 2 reflect that the sample comes from a waste dump of the Barruecopardo mine, and therefore from an area with low W contents.

The probability of fracture (S_i) and the kinetic parameters (α, α_T), a size of 80% of product undersize (P_{80}), and the reduction ratio were obtained from this procedure (R_r):

Figure 2 shows the values (S_i) plotted to the particle size according to the mill speed.

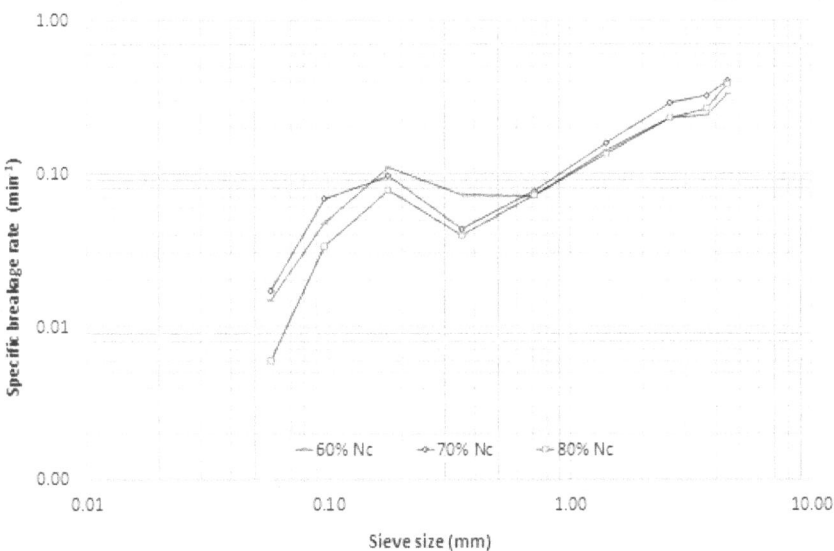

Figure 2. Evolution of the probability of fracture (S_i) with particle size, in relation to mill speed.

Figure 2 shows that comparing the three speeds tested, the highest probability of fracture occurred at 70% critical speed, agreeing with Gupta amd Sharma [9], Gupta [10], and Herbst and Fuerstenau [30], who did their experiments at lab scale, with several materials, simulations, and finally upscaling. This also verified the Steiner [29] recommendation of a configuration with F_r value close to 0.5. For intermediate and coarse grain sizes, a rather linear trend was observed using a logarithmic scale. This coincided with Deniz [22] for this grain size interval.

Noteworthy was a sharp break in slope that occurred at around 250 μm for the three tested speeds, coinciding with the size fractions more enriched in W, as shown in Table 2. Moreover, the results backed Gupta and Sharma's [9] statements, pointing to the probability of fracture S_i being one of the mill operational conditions more influenced by the mineralogical variability among monosize fractions.

On the other side, the evolution of S_i, P_{80}, and R_r for each feed monosize was studied at different mill speeds as summarized in Table 3. The reduction ratio was the result of dividing the d_{80} in the feed size (F_{80}) by the d_{80} in the product size (P_{80}).

Table 3. S_i, R_r, and P_{80} values as a function of working speed for each monosize.

Monosize (μm)	Specific Rate of Breakage, S_i (1/min)			Reduction Ratio, R_r			P_{80} (μm)		
	60% Nc	70% Nc	80% Nc	60% Nc	70% Nc	80% Nc	60% Nc	70% Nc	80% Nc
5000/4000	0.331	0.403	0.383	2.14	2.59	2.38	2244.9	1856.6	2014.7
4000/3350	0.241	0.322	0.266	1.66	2.08	1.74	2329.4	1864.2	2220.5
3350/2000	0.231	0.289	0.232	1.92	2.13	1.85	1608.1	1447.7	1665.0
2000/1000	0.143	0.159	0.135	1.67	1.84	1.69	1078.5	977.2	1063.8
1000/500	0.071	0.077	0.072	1.13	1.14	1.13	792.9	786.1	799.7
500/250	0.073	0.043	0.039	1.11	1.06	1.05	406.0	425.8	427.9
250/125	0.110	0.097	0.078	1.18	1.17	1.10	191.1	192.3	204.6
125/75	0.047	0.068	0.033	1.04	1.07	1.03	110.2	107.8	111.9
75/45	0.015	0.017	0.006	1.01	1.01	1.01	68.4	68.1	68.2

It must be highlighted in Table 3 that once the total grinding time was reached, a finer P_{80} and a higher R_r, were obtained at 70% working speed for most of the monosizes. This confirmed that the mill speed affected the grinding product [11,31].

The kinetic parameters (α, α_T) obtained after linearization of Equation (4) are summarized in Table 4 and the selection function is represented in Figure 3.

Table 4. Values of the kinetic parameters α, α_T for different mill speeds.

	60% Nc	70% Nc	80% Nc
α	0.42	0.58	0.74
α_T (1/min)	0.14	0.14	0.11

Table 4 shows that the parameter α is coherent with what Austin et al. [25] reported. These authors pointed out that it usually ranges between 0.5 and 1.5 and that it depends only on the mineral. On the other side, Table 4 and Figure 3 show that the value of the selection function α_T does not vary significantly despite the increasing speed, because the mill geometry remains unchanged. Figure 4 displays α_T values calculated from Equation (9), as proposed by Austin and Brame [28]. It can be seen that this expression does not fit this case. This led to a polynomial adjustment using the experiment values, as it was shown in Equation (11) that it fitted better with the studied sample.

$$\alpha_T = -1.775\, v_c^2 + 2.3625 v_c - 0.6402 \qquad (11)$$

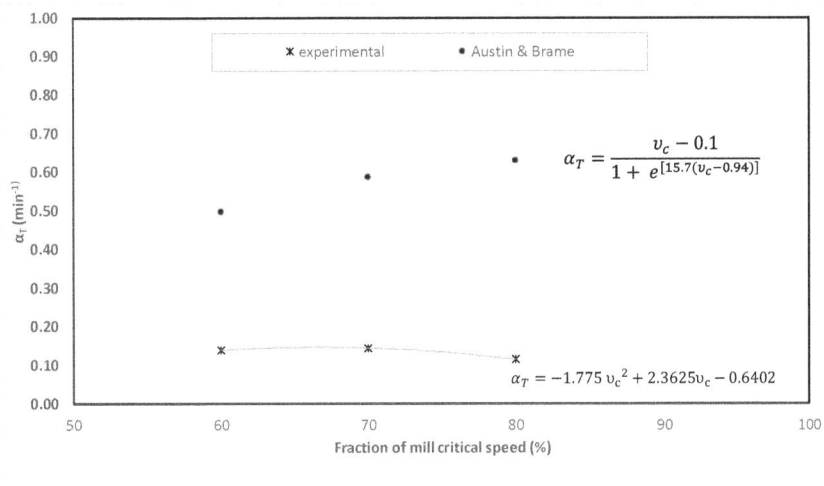

Figure 3. Variation of α_T with the working speed fraction.

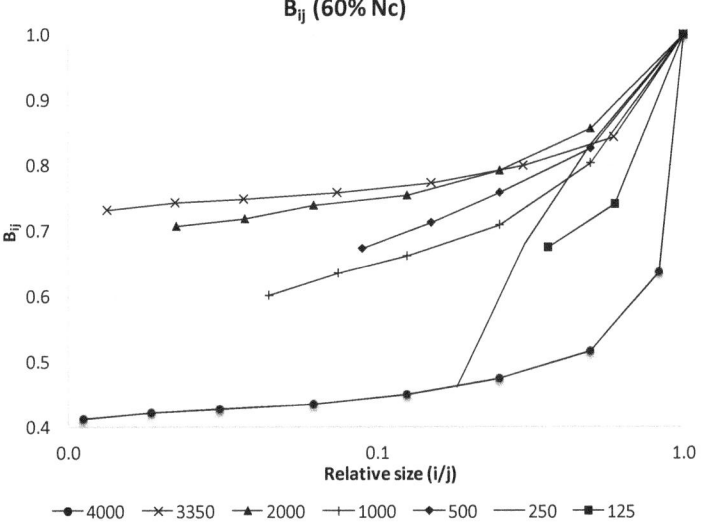

Figure 4. Behavior of the fracture function (B_{ij}), with respect to particle size. 60% working speed Nc (Barruecopardo ore).

4.2. Fracture Function, Kinetic Parameters (ϕ_j, γ and β)

The values of fracture function B_{ij} in relation to the particle size for each monosize as the mill speed varied were determined through Equation (5) and are shown in Figures 4–6 for 60%, 70%, and 80% mill speed, respectively.

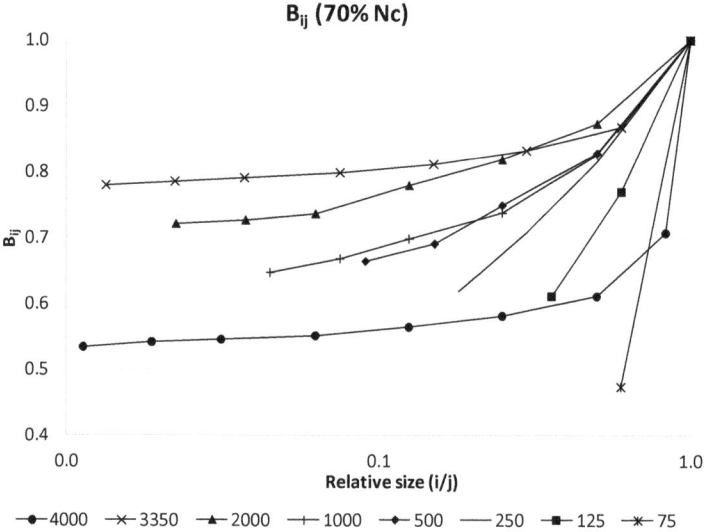

Figure 5. Behavior of the fracture function (B_{ij}), with respect to particle size. 70% working speed Nc (Barruecopardo ore).

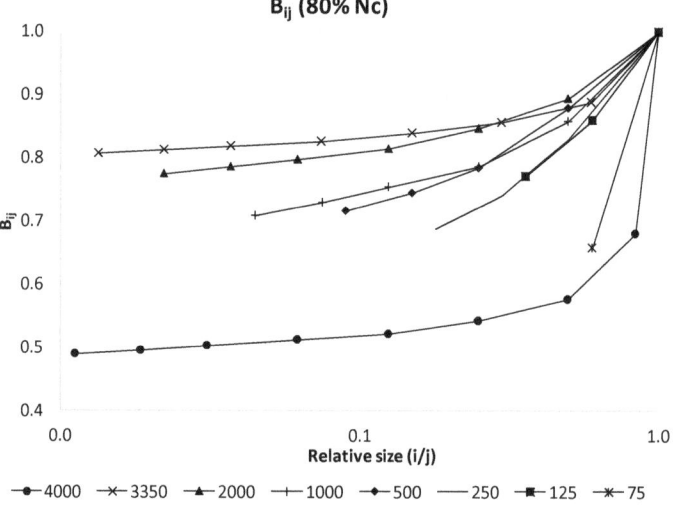

Figure 6. Behavior of the fracture function (B_{ij}), with respect to particle size. 80% working speed Nc (Barruecopardo ore).

The kinetic parameters of fracture function (ϕ_j, γ, and β) are shown in Table 5. According to Austin et al. (1984), ϕ_j and β are parameters that depend on the material. Regarding γ and β, these authors propose that their values were usually in the range of 0.5–1.5 and 2.5–5.0, respectively.

Table 5. Values of the kinetic parameters (φ_j, γ, β) for each monosize and for different mill speed conditions.

Monosize (μm)	j	j/i	B_{ij} (Test)	60% Nc φ_j	γ	β	B_{ij} (Test)	70% Nc φ_j	γ	β	B_{ij} (Test)	80% Nc φ_j	γ	β
4000	4000	1.000	1.000	0.483	0.036	3.285	1.000	0.586	0.021	3.361	1.000	0.552	0.027	3.477
	3350	0.838	0.637				0.708				0.680			
	2000	0.500	0.516				0.612				0.576			
	1000	0.250	0.474				0.581				0.542			
	500	0.125	0.450				0.563				0.522			
	250	0.063	0.435				0.551				0.512			
	125	0.031	0.426				0.545				0.503			
	75	0.019	0.421				0.540				0.495			
	45	0.011	0.411				0.534				0.489			
3350	3350	1.000	1.000	0.805	0.022	2.732	1.000	0.828	0.015	2.507	1.000	0.861	0.015	2.734
	2000	0.597	0.844				0.869				0.888			
	1000	0.299	0.799				0.833				0.856			
	500	0.149	0.774				0.812				0.838			
	250	0.075	0.758				0.799				0.825			
	125	0.037	0.749				0.791				0.818			
	75	0.022	0.743				0.785				0.812			
	45	0.013	0.731				0.779				0.807			
2000	2000	1	1	0.840	0.047	1.905	1	0.874	0.055	3.123	1	0.882	0.035	1.862
	1000	0.5	0.856				0.875				0.893			
	500	0.25	0.793				0.818				0.845			
	250	0.125	0.755				0.779				0.814			
	125	0.062	0.738				0.736				0.798			
	75	0.037	0.719				0.726				0.786			
	45	0.022	0.707				0.721				0.774			
1000	1000	1	1	0.807	0.094	2.048	1.000	0.821	0.078	1.885	1.000	0.855	0.061	1.939
	500	0.5	0.803				0.826				0.857			
	250	0.25	0.708				0.738				0.785			
	125	0.125	0.662				0.698				0.753			
	75	0.075	0.635				0.668				0.729			
	45	0.045	0.601				0.647				0.707			
500	500	1	1	0.887	0.115	4.054	1.000	0.876	0.118	2.538	1.000	0.882	0.087	1.283
	250	0.5	0.826				0.829				0.879			
	125	0.25	0.757				0.750				0.783			
	75	0.15	0.712				0.691				0.743			
	45	0.09	0.673				0.665				0.716			

Figures 4–6 and Table 5 show that B_{ij} depends on the feed grain size for parameters of 60%, 70%, and 80% of critical speed. The influence that mill speed exerts on B_{ij}, can also be noticed by comparing the different monosizes: a greater difference existed for coarser sizes, whereas it was lesser for finer sizes. This was due to the fact that coarse sizes not only possessed a higher S_i, but also were more prone to yield new finer particles (progeny). That meant that B_{ij} depended on the feed particle size, as Ipek and Goktepe [32] observed, which was also influenced by the mill speed, and concurred with results by Deniz [22]. Nevertheless, this variation was not as significant as reported by Austin et al. [25].

Table 5 shows parameter γ, which represents the fineness factor. In Figure 7, the γ values are plotted against mill speed for two feed particle sizes (4000 and 500 μm, respectively).

Figure 7 depicts that γ values are influenced by both mill speed and feed particle size. Smaller γ values were related to coarse particles (4000 μm), which meant that more fine particles were generated. Conversely, finer particles (500 μm) generated a lesser proportion of fine particles, agreeing well with results by Ipek and Goktepe [32] and Austin et al. [25].

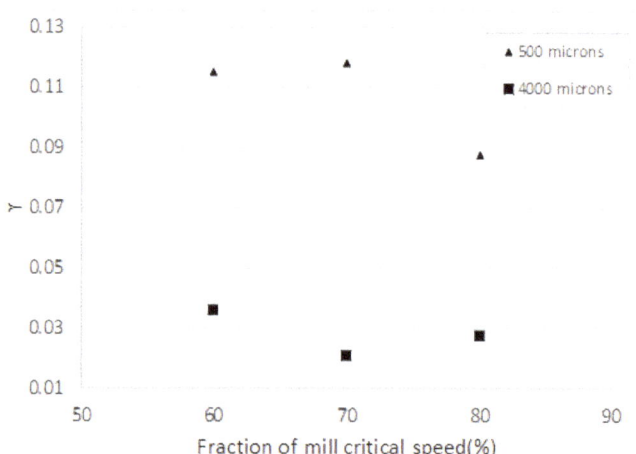

Figure 7. Variation of γ with mill speed.

4.3. Chemical Characterization of the Grinding Products

Figures 8–10 illustrate the evolution of W content in the product in relation to the feed monosizes and their grain size fractions for each mill speed.

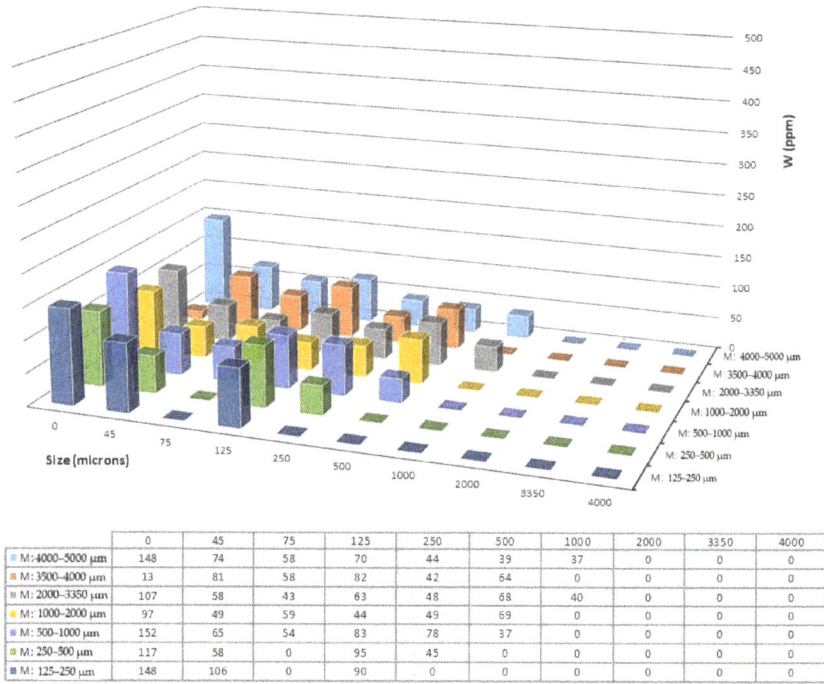

	0	45	75	125	250	500	1000	2000	3350	4000
M: 4000–5000 μm	148	74	58	70	44	39	37	0	0	0
M: 3500–4000 μm	13	81	58	82	42	64	0	0	0	0
M: 2000–3350 μm	107	58	43	63	48	68	40	0	0	0
M: 1000–2000 μm	97	49	59	44	49	69	0	0	0	0
M: 500–1000 μm	152	65	54	83	78	37	0	0	0	0
M: 250–500 μm	117	58	0	95	45	0	0	0	0	0
M: 125–250 μm	148	106	0	90	0	0	0	0	0	0

Figure 8. Evolution of the W grade in the product with respect to the monosizes and their grain size fractions for 60% Nc.

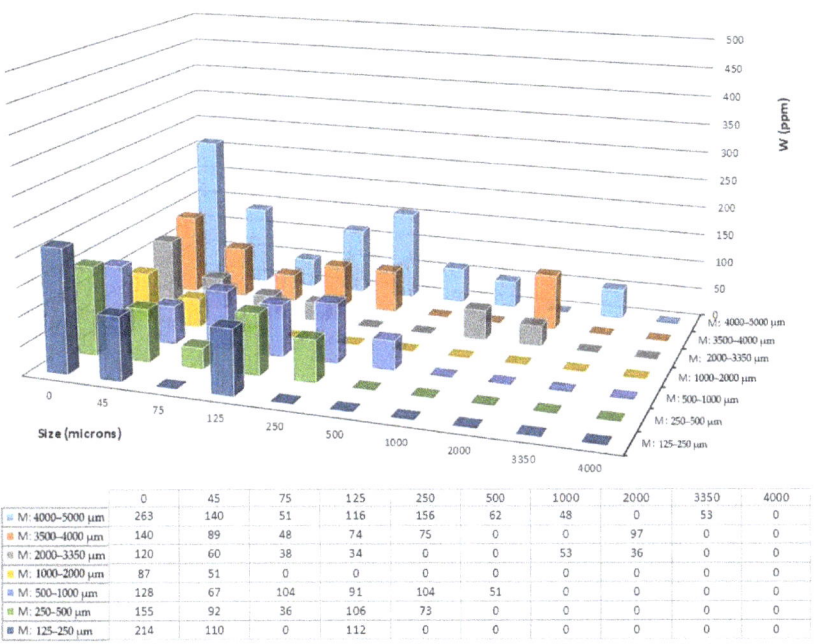

	0	45	75	125	250	500	1000	2000	3350	4000
M: 4000–5000 μm	263	140	51	116	156	62	48	0	53	0
M: 3500–4000 μm	140	89	48	74	75	0	0	97	0	0
M: 2000–3350 μm	120	60	38	34	0	0	53	36	0	0
M: 1000–2000 μm	87	51	0	0	0	0	0	0	0	0
M: 500–1000 μm	128	67	104	91	104	51	0	0	0	0
M: 250–500 μm	155	92	36	106	73	0	0	0	0	0
M: 125–250 μm	214	110	0	112	0	0	0	0	0	0

Figure 9. Evolution of the W grade in the product with respect to the monosizes and their grain size fractions for 70% *Nc*.

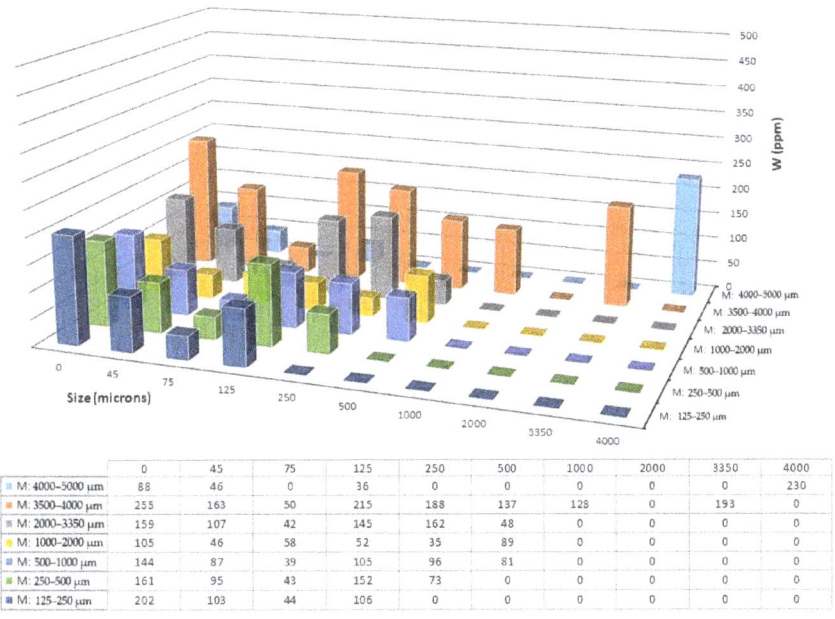

	0	45	75	125	250	500	1000	2000	3350	4000
M: 4000–5000 μm	88	46	0	36	0	0	0	0	0	230
M: 3500–4000 μm	255	163	50	215	188	137	128	0	193	0
M: 2000–3350 μm	159	107	42	145	162	48	0	0	0	0
M: 1000–2000 μm	105	46	58	52	35	89	0	0	0	0
M: 500–1000 μm	144	87	39	105	96	81	0	0	0	0
M: 250–500 μm	161	95	43	152	73	0	0	0	0	0
M: 125–250 μm	202	103	44	106	0	0	0	0	0	0

Figure 10. Evolution of the W grade in the product with respect to the monosizes and their grain size fractions for 80% *Nc*.

As can be seen in Figures 8–10, for most monosizes, W content increased with decreasing particle size. This could be partly explained because feed grain sizes around 250 μm already yielded higher W content, as shown in Table 2. Nevertheless, maximum values after grinding were 6–7 times higher than original values pointing undoubtedly to a differential grinding effect leading to W mineral grains mainly falling in the 250–125 μm interval. The accumulation of these W-enriched particles, which were more difficult to grind, supports the mineralogical explanation of the aforementioned breakage probability reduction at 250 μm size. In addition, this interval always presented S_i values higher than the coarser and finer intervals. This would suggest that particles of this size would have a higher probability of fracture compared to the adjacent size intervals. Indeed, in all cases, a decrease in W content could be observed down to 75 μm size, followed by an increase of further finer particles. It must be highlighted that the highest W content was yielded for grinding at 70% Nc. This could be explained because, under these grinding conditions, the mill performance was more efficient due to a more adequate charge regime (F_r close to 0.5), leading to better grinding kinetics.

5. Conclusions

The experimental work here presented and its further analysis permits to draw the following conclusions:

- Austin's methodology has allowed us to characterize the kinetic behavior of tungsten ore by determining the kinetic parameters (α, α_T, φ_j, γ, β), concluding that the values S_i and B_{ij} do not vary significantly with time.
- The probability of fracture, S_i, is highest at 70% critical speed. Fracture function B_{ij}, does not vary significantly with mill speed. Nevertheless, it is affected by the feed particle size, becoming higher for coarser sizes.
- Equation (10) is proposed as the best α_T fit, specifically for the studied ore.
- Values of parameter γ are influenced by both mill speed and feed particle size. Coarser particles yield smaller γ values, i.e., they produce more fines, whereas finer particles produce lesser quantities. The parameters ϕ_j and β depend on the features of the ore.
- The chemical characterization and the study of the evolution of the relevant grinding parameters, such as P_{80} and R_r, to the grinding time have demonstrated that, first, the highest probability of fracture occurs at 70% of the critical speed and; second, the effect of differential grinding is evidenced between the W-bearing species and the gangue. The latter results in an increase of the W grade in the monosize 250/125 μm.

Author Contributions: Conceptualization and execution of experiments, J.V.N. and J.M.M.-A.; methodology and investigation, J.V.N. and A.L.C.-V.; formal analysis and data curation, J.V.N. and J.M.M.-A.; writing–original draft preparation, J.V.N.; writing–review and editing, J.V.N. and J.M.M.-A.; supervision, J.M.M.-A. and A.L.C.-V.; project administration and funding acquisition, J.M.M.-A. All authors have read and agreed to the submitted version of the manuscript.

Funding: This work is part of the OPTIMORE project funded by the European Union Horizon 2020 Research and Innovation Programme under grant agreement No 642201.

Institutional Review Board Statement: Not applicable.

Informed Consent Statement: Not applicable.

Data Availability Statement: No new data were created or analyzed in this study. Data sharing is not applicable to this article.

Conflicts of Interest: The authors declare no conflict of interest.

References

1. European Commission. Study on the EU's List of Critical Raw Materials. 2020. Available online: https://ec.europa.eu/commission/presscorner/detail/en/ip_20_1542 (accessed on 25 November 2020).
2. Roskill. Tungsten. *Market Report*. 2020. Available online: https://roskill.com/market-report/tungsten/ (accessed on 25 November 2020).

3. Schmidt, S. *From Deposit to Concentrate: The Basics of Tungsten Mining Part 2: Operational Practices and Challenges*; International Tungsten Industry Association: London, UK, 2012; pp. 1–20. Available online: http://www.itia.info/assets/files/newsletters/Newsletter_2012_06.pdf (accessed on 25 November 2020).
4. Ormonde Mining PLC. *Annual Report & Accounts 2018*; Issue 256353; Ormonde Mining PLC Co.: Meath, Ireland, 2018.
5. Murciego, A.; Álvarez-Ayuso, E.; Pellitero, E.; Rodríguez, M.A.; García-Sánchez, A.; Tamayo, A.; Rubio, J.; Rubio, F.; Rubin, J. Study of arsenopyrite weathering products in mine wastes from abandoned tungsten and tin exploitations. *J. Hazard. Mater.* **2011**, *186*, 590–601. [CrossRef] [PubMed]
6. Alfonso, P.; Castro, D.; Garcia-Valles, M.; Tarragó, M.; Tomasa, O.; Martínez, S. Recycling of tailings from the Barruecopardo tungsten deposit for the production of glass. *J. Therm. Anal. Calorim.* **2016**, *125*, 681–687. [CrossRef]
7. Fuerstenau, D.W.; Lutch, A.D.J.J. The effect of ball size on the energy efficiency of hybrid high-pressure roll mill/ball mill grinding. *Powder Technol.* **1999**, *105*, 199–204. [CrossRef]
8. Menéndez-Aguado, J.M.; Velázquez, A.L.C.; Tijonov, O.N.; Díaz, M.A.R. Implementation of energy sustainability concepts during the comminution process of the Punta Gorda nickel ore plant (Cuba). *Powder Technol.* **2006**, *170*, 153–157. [CrossRef]
9. Gupta, V.K.; Sharma, S. Analysis of ball mill grinding operation using mill power specific kinetic parameters. *Adv. Powder Technol.* **2014**, *25*, 625–634. [CrossRef]
10. Gupta, V.K. Determination of the specific breakage rate parameters using the top-size-fraction method: Preparation of the feed charge and design of experiments. *Adv. Powder Technol.* **2016**, *27*, 1710–1718. [CrossRef]
11. Öksüzoglu, B.; Uçurum, M. An experimental study on the ultra-fine grinding of gypsum ore in a dry ball mill. *Powder Technol.* **2016**, *291*, 186–192. [CrossRef]
12. Pedrayes, F.; Norniella, J.G.; Melero, M.G.; Menéndez-Aguado, J.M.; del Coz-Díaz, J.J. Frequency domain characterization of torque in tumbling ball mills using DEM modelling: Application to filling level monitoring. *Powder Technol.* **2018**, *323*, 433–444. [CrossRef]
13. Deniz, V. Influence of interstitial filling on breakage kinetics of gypsum in ball mill. *Adv. Powder Technol.* **2011**, *22*, 512–517. [CrossRef]
14. Chimwani, N.; Glasser, D.; Hildebrandt, D.; Metzger, M.J.; Mulenga, F.K. Determination of the milling parameters of a platinum group minerals ore to optimize product size distribution for flotation purposes. *Miner. Eng.* **2013**, *43–44*, 67–78. [CrossRef]
15. Lund, C.; Lamberg, P. Geometallurgy-A tool for better resource efficiency. *Met. Miner.* **2014**, *37*, 39–43.
16. Mwanga, A.-R. Development of a Geometallurgical Testing Framework for Ore Grinding and Liberation Properties. Ph.D. Thesis, Lulea University of Technology, Lulea, Sweden, 2016.
17. Mwanga, A.; Parian, M.; Lamberg, P.; Rosenkranz, J. Comminution modeling using mineralogical properties of iron ores. *Miner. Eng.* **2017**, *111*, 182–197. [CrossRef]
18. Little, L.; Mainza, A.N.; Becker, M.; Wiese, J. Fine grinding: How mill type affects particle shape characteristics and mineral liberation. *Miner. Eng.* **2017**, *111*, 148–157. [CrossRef]
19. Gupta, V.K. Effect of size distribution of the particulate material on the specific breakage rate of particles in dry ball milling. *Powder Technol.* **2017**, *305*, 714–722. [CrossRef]
20. Ciribeni, V.; Bertero, R.; Tello, A.; Puerta, M.; Avellá, E.; Paez, M.; Aguado, J.M.M. Application of the Cumulative Kinetic Model in the Comminution of Critical Metal Ores. *Metals* **2020**, *10*, 925. [CrossRef]
21. Hesse, M.; Popov, O.; Lieberwirth, H. Increasing efficiency by selective comminution. *Miner. Eng.* **2017**, *103–104*, 112–126. [CrossRef]
22. Deniz, V. The effect of mill speed on kinetic breakage parameters of clinker and limestone. *Cem. Concr. Res.* **2004**, *34*, 1365–1371. [CrossRef]
23. Petrakis, E.; Komnitsas, K. Improved Modeling of the Grinding Process through the Combined Use of Matrix and Population Balance Models. *Minerals* **2017**, *7*, 67. [CrossRef]
24. Austin, L. A Review—Introduction to the mathematical description of grinding as a rate process. *Powder Technol.* **1972**, *5*, 1–17. [CrossRef]
25. Austin, L.G.; Klimpel, R.R.; Luckie, P.T. *Process Engineering of Size Reduction: Ball Milling*; SME—AIME: New York, NY, USA, 1984.
26. Wang, X.; Gui, W.; Yang, C.; Wang, Y. Wet grindability of an industrial ore and its breakage parameters estimation using population balances. *Int. J. Miner. Process.* **2011**, *98*, 113–117. [CrossRef]
27. Petrakis, E.; Stamboliadis, E.; Komnitsas, K. Identification of Optimal Mill Operating Parameters during Grinding of Quartz with the Use of Population Balance Modeling. *KONA Powder Part. J.* **2017**, *34*, 213–223. [CrossRef]
28. Austin, L.G.; Brame, K. A comparison of the Bond method for sizing wet tumbling ball mills with a size—mass balance simulation model. *Powder Technol.* **1983**, *34*, 261–274. [CrossRef]
29. Steiner. Characterization of laboratory-scale tumbling mills. *Int. J. Miner. Process.* **1996**, *44–45*, 373–382.
30. Herbst, J.A.; Fuerstenau, D.W. Mathematical simulation of dry ball milling using specific power information. *Trans. AIME* **1973**, *254*, 343–348.
31. Cayirli, S. Influences of operating parameters on dry ball mill performance. *Physicochem. Probl. Miner. Process.* **2018**, *54*, 751–762.
32. Ipek, H.; Goktepe, F. Determination of grindability characteristics of zeolite. *Physicochem. Probl. Miner. Process.* **2011**, *47*, 183–192.

Article

Variability Study of Bond Work Index and Grindability Index on Various Critical Metal Ores

Gloria G. García [1], Josep Oliva [2], Eduard Guasch [2], Hernán Anticoi [3], Alfredo L. Coello-Velázquez [4] and Juan M. Menéndez-Aguado [1,*]

1. Escuela Politécnica de Mieres, University of Oviedo, Gonzalo Gutiérrez Quirós, 33600 Mieres, Spain; gloria.glez.gcia@gmail.com or UO150088@uniovi.es
2. Departament D'Enginyeria Minera, Industrial i TIC, Universitat Politècnica De Catalunya Barcelona Tech, Av. Bases De Manresa, 08242 Manresa, Spain; josep.oliva@upc.edu (J.O.); eduard.guasch@upc.edu (E.G.)
3. Escuela Politecnica de Ingenieria de Minas y Energia, Universidad de Cantabria, Ronda Rufino Peón, 39316 Torrrelavega, Spain; hernan.anticoi@unican.es
4. CETAM, Universidad de Moa Dr. Antonio Núñez Jiménez, Moa 83300, Cuba; acoello@ismm.edu.cu
* Correspondence: maguado@uniovi.es; Tel.: +34-985458033

Abstract: It is a well-known fact that the value of the Bond work index (w_i) for a given ore varies along with the grinding size. In this study, a variability bysis is carried out with the Bond standard grindability tests on different critical metal ores (W, Ta), ranging from coarse grinding (rod mills) to fine grinding (ball mills). The relationship between w_i and grinding size did not show a clear correlation, while the grindability index (gpr) and the grinding size showed a robust correlation, fitting in all cases to a quadratic curve with a very high regression coefficient. This result suggests that, when performing correlation studies among ore grindability and rock mechanics parameters, it is advised to use the grindability index instead of the Bond work index.

Keywords: grindability; comminution; Bond work index

Citation: García, G.G.; Oliva, J.; Guasch, E.; Anticoi, H.; Coello-Velázquez, A.L.; Menéndez-Aguado, J.M. Variability Study of Bond Work Index and Grindability Index on Various Critical Metal Ores. *Metals* 2021, 11, 970. https://doi.org/10.3390/met11060970

Academic Editor: Jean François Blais

Received: 14 May 2021
Accepted: 14 June 2021
Published: 17 June 2021

Publisher's Note: MDPI stays neutral with regard to jurisdictional claims in published maps and institutional affiliations.

Copyright: © 2021 by the authors. Licensee MDPI, Basel, Switzerland. This article is an open access article distributed under the terms and conditions of the Creative Commons Attribution (CC BY) license (https://creativecommons.org/licenses/by/4.0/).

1. Introduction

Comminution is an essential operation for the mining and mineral processing industry. It also plays a central role in the cement production, ceramics and chemical industries. In the mineral industry, the liberation of valuable minerals from the gangue is a fundamental requirement for all subsequent separation or extraction operations, and this is achieved through several stages of rock fragmentation, that is, by comminution of the ore [1].

Schönert [2] estimated that minerals comminution consumes 3% of all the energy produced by industrialized countries, in line with former studies [3]. More recent evaluations estimate that comminution operations are responsible for 3–5% of energy consumption at a global scale [4]. Moreover, in terms of OPEX in mineral processing plants, comminution operations amount to 40–50% of the energy consumption.

Considering the above, any gain in efficiency can significantly impact the plant operating costs and the consequent conservation of resources [5]. In this sense, an improvement in knowledge of ore grinding behavior can allow modification of the operation and control strategies of the grinding operations, resulting in significant energy savings. This would increase the competitiveness of operations and reduce emissions.

It is common to process multicomponent ores, made up of at least two mineralogical components with differences in their physical and physicomechanical properties. Some authors [6,7] show that disregarding the variability of the feed mineralogical composition produces large deviations in the planned metallurgical efficiency, along with problems in the treatment of the ores with such characteristics. On the other hand, between the initial exploration work for the design of any mineral beneficiation plant and the reaching of its full operating regime, and even after reaching it, there will be variations in the plant feed composition, implying substantial changes in mineral properties. Therefore, it would be advisable to adjust the operating and control conditions of the treatment plant in general and the size reduction section in particular.

The energy–size relationships in comminution processes have been an object of research since the first industrial revolution [8]. Rittinger [9] proposed the first law of comminution, supposing that the amount of created surface is proportional to the specific energy consumption in grinding operations, as expressed in Equation (1):

$$E = K_R \cdot \left(\frac{1}{P} - \frac{1}{F}\right) \tag{1}$$

where E is the specific energy consumption [kWh/t], K_R is the proportionality coefficient and P and F are the particle sizes of the product and feed, respectively [μm].

Kick [10], in the second law of comminution, argued that according to his calculations, the specific energy consumption would be proportional to the volume reduction, as expressed in Equation (2), where K_K is a different proportionality coefficient.

$$E = K_K \cdot \left(\frac{1}{\ln(P)} - \frac{1}{\ln(F)}\right) \tag{2}$$

The differences between the Rittinger and Kick models lasted for years, until the proposal of the third theory of comminution by F. Bond [11–13], which is summarized in Equation (3),

$$E = K_B \cdot \left(\frac{1}{\sqrt{P}} - \frac{1}{\sqrt{F}}\right) \tag{3}$$

where $K_B = 10 \cdot w_i$, and w_i is expressed in kWh/t.

Subsequent studies [14] explained that the three laws derive from a generalized comminution differential equation, each one best applied to a different size range (Rittinger's law for fine grinding; Bond's law for coarse grinding and secondary/tertiary crushing; and Kick's law for primary crushing). The novelty in the third law's proposal was the procedure for determining w_i in the case of crushing, rod milling and ball milling [13,15]. The practical interest of w_i is unquestionable. From a technical perspective, it constitutes the most reliable method of characterizing ore grindability when designing the necessary tumbling mills to process that ore. Bearman et al. (1997) showed that other mechanical characterization tests are insufficient when predicting the grinding ore behavior.

A logical reasoning process should suggest finding some correlation among mechanical parameters (hardness, Young's modulus, uniaxial compression strength (UCS), etc.) and the ore grinding behavior. Several researchers [4,16–18] followed that inspiration, but no generalizable results have been obtained since grindability behavior is usually evaluated under closed-circuit conditions, which means that not only breakage but breakage plus classification operations are involved. Moreover, we can easily find ores with high hardness and high grindability values, but among the highest grindability values, we can find quite soft ores (graphite or mica group minerals). On the other hand, diamond mineral shows modest grindability values. Thus, it is worth emphasizing that the Bond work index tries to characterize the ore grinding behavior in a closed circuit, encompassing the ore mechanical behavior before the mill action (i.e., whatever the type of the mill and its characteristics of action), but also the screening or classification stage involved in the closed circuit, which is greatly influenced by product size and shape.

Due to the fact that Bond's proposal was undoubtedly linked to a market-dominant firm, i.e., Allis Chalmers, which even owned the patent of the standard mill, several proposals soon emerged to define alternative test approaches, which can be grouped in the following types:

- Indirect work index determination in other lab mills [19–24].
- Specific energy determination from correlations in different devices [25–27].
- Work index calculation through lab tests and simulation [28–31].

It is essential to notice that despite the almost unanimous consideration of the w_i as the characteristic parameter of ore grinding behavior, it is not fully understood at the industrial level, even being handled as a constant value. Bond himself usually reported in his papers separately the grindability values for the Bond rod mill test (BRM) and the Bond ball mill test (BBM), but no study could be found analyzing the information from BRM and BBM test values and deepening them to explain the variability obtained.

In this work, the analysis of grindability results obtained in a broad particle size range and several critical metal ores (W, Ta) is carried out. The variability of the work indices in BRM and BBM tests is studied to propose a methodology to model said variability.

2. Materials and Methods

2.1. Materials

This study was carried out on three ores from W mines and two ores from a Ta mine. Two of the W ore samples were Scheelite ores, received from Barruecopardo (Spain) and Mittersill (Austria). A detailed description of Barruecopardo ore can be found in recent publications [32,33]. In the case of Mittersill, an ore description can be found in [34]. The third W ore was Wolframite from the Panasqueira Mine (Portugal), and a detailed description of this ore can be found in [35]. In the case of Ta ores, two different samples were received from Penouta Mine (Spain), one from the open pit and the other one from the tailings pond of the former Tin mining activities in that mine. Characterization studies of those samples have been previously published [36,37]. It must be pointed out that, in the particular case of Barruecopardo mine, two different samples were taken from different heaps. The sample size in each case, considering the largest particle size, was enough to perform the series of Bond ball mill grindability tests separately (see Section 2.2), but not enough to perform the series of Bond rod mill grindability tests separately for each sample (see Section 2.3). Accordingly, it was decided to blend and homogenize the Barruecopardo samples and perform the rod mill test on the samples blend.

2.2. Bond Ball Mill (BBM) Standard Test

The procedures to carry out the Bond grindability tests in ball mills and rod mills are outlined in Sections 2.2 and 2.3. They are usually referred to as the standard tests, but it must be highlighted that the procedures haven't been defined by ISO or ASTM standards. The closest attempt to a standard definition was the initiative of the Global Mining Standard Group [38].

The Bond work index most commonly referred to is the BBM work index. This value is obtained in a 12" × 12" laboratory mill running at 70 rpm, with rounded inner edges and without lifters. The grinding charge is comprised of a distribution of steel balls with several diameters. Table 1 shows the original Bond proposal [13], while the last Bond recommendation can be found in Table 2 [39].

Table 1. Ball grinding charge distribution proposed by Bond.

Ball Size		Balls	
Inch	cm	Number	Weight (g)
1.45	3.683	43	8803
1.17	2.972	67	7206
1.00	2.540	10	672
0.75	1.905	71	2011
0.61	1.549	94	1433
	Total:	285	20,125

Table 2. Ball charge distribution used in this research.

Ball Size		Balls	
Inch	cm	Number	Weight (g)
1.500	3.810	25	5690
1.25	3.175	39	5137
1.000	2.540	60	4046
0.875	2.223	68	3072
0.750	1.905	93	2646
	Total:	285	20,592

The mill feed must be prepared by controlled crushing until 100% passes through a 6 Tyler mesh (3.35 mm). The first grinding cycle feed must be 700 cm^3, and this volume's weight is fixed as the mill charge in all subsequent cycles. Additionally, fresh feed particle size distribution (PSD) is obtained to calculate the 80% passing size (F_{80}) and undersize weight already present in the feed.

The test procedure consists of performing several dry grinding cycles to simulate a continuous closed-circuit operation with 250% circulating load (Figure 1). The circuit is closed by a sieve (P_{100}) selected according to the industrial grinding size target, always between 28 and 325 Tyler mesh (40–600 microns).

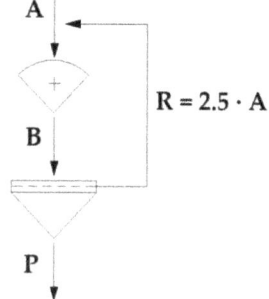

Figure 1. Closed-circuit BBM test objective layout.

The first cycle starts with an arbitrary number of mill revolutions, usually 100 revolutions with hard-to-grind ores and 50 revolutions with soft ores. The first run product is sieved, the undersize is weighed, and the net grams produced (gpr) of the first run is calculated, considering the undersize already present in the feed.

The second cycle feed is constituted by the former cycle's oversize product plus enough fresh feed to complete the initial 700 cm^3 weight. The second cycle number of revolutions is calculated considering the predefined circulating load value (250%), according to Equation (4),

$$n_i = \frac{(P_S - F_{f,i})}{gpr_{i-1}} \qquad (4)$$

where n_i is the number of mill revolutions at run i; P_S is the expected product weight once it reaches the steady state (g), calculated by dividing the initial 700 cm^3 weight by 3.5; $F_{f,i}$ is the weight of fines already in the feed (g), which can be calculated from the feed PSD and the total fresh feed weight added in the run i (which equals the total undersize product in the run $i-1$) and gpr_{i-1} is the net grams produced in the previous run, $i-1$.

Subsequent grinding cycles are carried out (at least five) until gpr reaches equilibrium. The final value of gpr is calculated as the average of the last three cycles. The final cycle product PSD is calculated to obtain P_{80}, and the BBM work index can be calculated using Equation (5),

$$w_i = \frac{44.5}{P_{100}^{0.23} \cdot gpr^{0.82} \cdot \left(\frac{10}{\sqrt{P_{80}}} - \frac{10}{\sqrt{F_{80}}}\right)} \quad (5)$$

where the BBM work index, w_i, is expressed in kWh/sht; P_{100}, F_{80} and P_{80} are expressed in microns and gpr is expressed in g/rev. Bond named gpr as the grindability index.

According to Bond [13], w_i should conform with the motor output power to an average overflow ball mill of 8 ft inner diameter grinding wet in a closed circuit. This value should be multiplied by correcting factors to conform with other situations, such as dry grinding (at least 1.30) or different inner mill diameters. A complete and updated description of correction factors was written by Rowland [40].

2.3. Bond Rod Mill (BRM) Standard Test

In this case, the procedure is very similar to BBM test, and only some differences are commented on [13]. The feed must be prepared until 100% passes $\frac{1}{2}''$ (1, 27 mm), with a feed volume of 1250 cm^3. Dry grinding cycles are performed with 100% circulating load in a laboratory rod mill 12" × 24" with a wave-type lining, running at 46 rpm. The grinding charge consists of six 1.25" diameter and two 1.75" diameter steel rods 21" long, weighing 33.380 kg. In this case, P_{100} values can range from 4 to 65 Tyler mesh (4.7 mm to 200 microns).

In order to equalize segregation at the mill ends, it is rotated level for eight revolutions, then tilted up 5° for one revolution, down 5° for another revolution and returned to level for eight revolutions continuously through each grinding cycle. At the end of each cycle, the mill is discharged by tilting downward at 45° for 30 revolutions. Once equilibrium is reached, gpr and P_{80} are calculated, and the BRM work index is calculated from Equation (6).

$$w_i = \frac{62}{P_{100}^{0.23} \cdot gpr^{0.625} \cdot \left(\frac{10}{\sqrt{P_{80}}} - \frac{10}{\sqrt{F_{80}}}\right)} \quad (6)$$

Again, w_i should conform with the motor output power to an average overflow rod mill of 8 ft inner diameter grinding wet in an open circuit.

2.4. Grindability Tests

A series of tests was defined to analyze the variation of grindability properties in the selected ores. Depending on sample availability, a minimum of three BBM tests and a minimum of 2 BRM tests were performed, each test at a different P_{100} for every ore. Then, the values of gbp and w_i were obtained for each ore, and an attempt to model their variation with P_{100} was performed in each case. Full details of the performed tests and results are available in the supplementary material.

It is generally accepted that, provided samples are representative, BBM and BRM grindability test repetitions are unnecessary. This is justified by the iterative nature of the grindability tests procedures, and both rod and ball mill tests' repeatability were proven to be less than ±4% at two standard deviations [41].

3. Results and Discussion

In the case of Penouta tailings pond ore, the variation of w_i versus P_{100} is plotted in Figure 2, for both BBM and BRM tests. The obtained values show a lack of continuity, and a clear trend function could hardly be defined. Nonetheless, when observing Figure 3, which depicts the variation of gpr versus P_{100} in both BBM and BRM tests, a fairly clear trend can be seen; according to this, Figure 3 also shows the quadratic fit of gpr consolidated values versus P_{100}, with a determination coefficient of 99.76%.

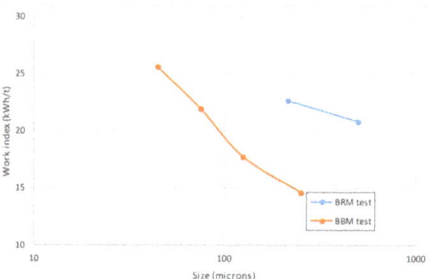

Figure 2. Variation of BBM and BRM w_i values with P_{100}, Penouta tailings pond ore.

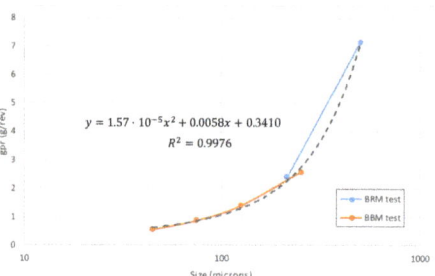

Figure 3. Variation of BBM and BRM *gpr* values with P_{100}, Penouta tailings pond ore.

A similar analysis was performed in the case of Penouta mine ore (see Figures 4 and 5). In this case, despite w_i versus P_{100} plot revealing a lack of continuity again (Figure 4), plotting *gpr* versus P_{100} (Figure 5) showed a similar trend to the previous ore. Moreover, the quadratic fit was almost perfect in this case, with a coefficient of determination of 100.00%.

In the case of Mittersill ore (Figures 6 and 7), the transition between BBM and BRM w_i values with P_{100} shows a better continuity than in previous cases (Figure 7), so the determination coefficient reached again a very high value, 99.89%.

Plotting BBM and BRM w_i values versus P_{100} in the case of Panasqueira ore yeilded a clear trend in the case of BBM w_i values, but a with a roller-coaster type shape in the case of BRM w_i values (Figure 8). Unexpectedly, when plotting *gpr* values versus P_{100} (Figure 9), again a quadratic fit yielded a very high value of the determination coefficient, 99.10%.

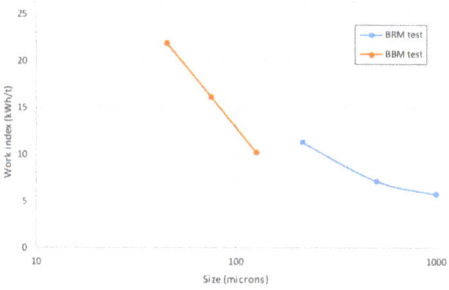

Figure 4. Variation of BBM and BRM w_i values with P_{100}, Penouta mine ore.

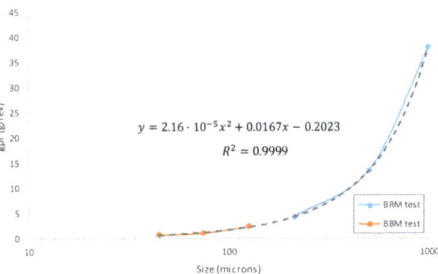

Figure 5. Variation of BBM and BRM *gpr* values with P_{100}, Penouta mine ore.

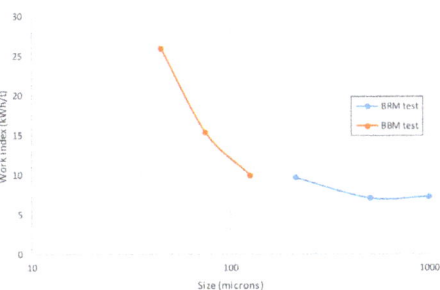

Figure 6. Variation of BBM and BRM w_i values with P_{100}, Mittersill ore.

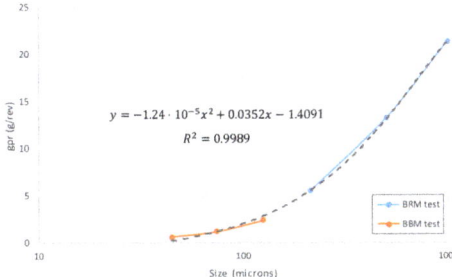

Figure 7. Variation of BBM and BRM *gpr* values with P_{100}, Mittersill ore.

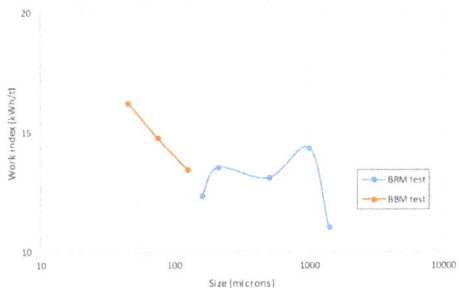

Figure 8. Variation of BBM and BRM w_i values with P_{100}, Panasqueira ore.

Finally, Barruecopardo ore samples results are depicted in Figures 10 and 11. As mentioned above, BBM tests were performed on the same ore samples but with different origins, while BRM tests were performed on the composite obtained after blending both samples. Once more, with an evident lack of continuity in the w_i versus P_{100} plot (Figure 10), a clear quadratic trend was obtained when plotting gpr values versus P_{100} (Figure 11), with a very high value of the determination coefficient, 99.95%.

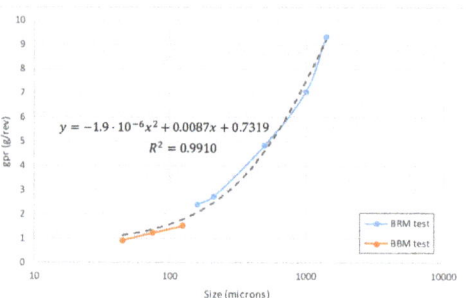

Figure 9. Variation of BBM and BRM gpr values with P_{100}, Panasqueira ore.

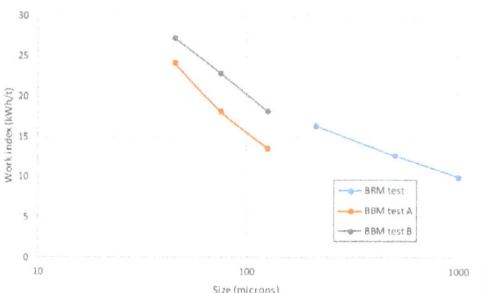

Figure 10. Variation of BBM and BRM w_i values with P_{100}, Barruecopardo ore.

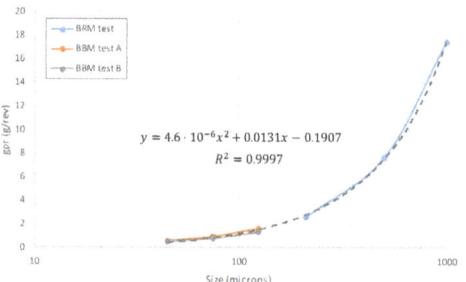

Figure 11. Variation of BBM and BRM gpr values with P_{100}, Barruecopardo ore.

Given the results obtained, it is evident that there is significant variability of w_i values with grinding size, both in BBM and BRM grindability tests. While w_i versus P_{100} plots show no continuity in general (being erratic in the case of Panasqueira ore, BRM w_i values) when plotting gpr versus P_{100}, a parabolic shape is clearly depicted with all ores. Furthermore, the quadratic fitting determination coefficients overcame 99.7% in all cases.

A recommendation can be made in the light of these results: any energy consumption model based on correlating w_i with mechanical parameters (geotechnical) or operational parameters (drilling, blasting) should be revised considering gpr values instead of w_i values, which probably would yield a better determination coefficient.

These results also invited us to perform a conceptual review of the Fred Bond literature to seek relevant considerations about the significance of gpr. Thus, it gains additional value that gpr was already named "grindability" since the paper led by Walter Maxson [42], in which Fred Bond was also a co-author. Fred Bond, in his subsequent papers, also utilized this definition. Considering that w_i is worldwide known as the Bond index, and without the intention of subtracting an iota of importance from the broad contribution of Fred Bond (w_i is the most practical tool in rod and ball mill calculation), it seems fair to propose the naming of gpr as the *Maxson index*. This so-called Maxson index should be meaningful, not only for being the critical parameter to obtain the Bond work index but also for characterizing the ore breakage behavior.

4. Conclusions

The following conclusions summarize the results obtained in this research:

- According to the obtained results, BBM and BRM grindability tests showed no continuity or clear correlation when considering w_i values versus P_{100}, but a clear tendency was obtained in all cases when plotting gpr versus P_{100}.
- It is advised that energy consumption modelling based on correlations involving w_i and other mechanical or operational parameters would yield a better determination coefficient using gpr values instead.
- The re-signifying of gpr evidenced to characterize the ore breakage behavior and its origin justify the proposal of naming gpr as the Maxson grindability index.

Supplementary Materials: The following are available online at https://www.mdpi.com/article/10.3390/met11060970/s1, Table S1. PENOUTA (tailings pond) BRM, Table S2. PENOUTA (tailings pond) BBM, Table S3. PENOUTA (mine) BRM, Table S4. PENOUTA (mine) BBM, Table S5. MITTERSILL BRM, Table S6. MITTERSILL BBM, Table S7. PANASQUEIRA BRM, Table S8. PANASQUEIRA BBM, Table S9. BARRUECOPARDO BRM, Table S10. BARRUECOPARDO BBM (test A), Table S11. BARRUECOPARDO BBM (test B).

Author Contributions: Conceptualization, A.L.C.-V. and J.M.M.-A.; methodology, G.G.G. and J.M.M.-A.; validation, G.G.G. and J.M.M.-A.; formal analysis, G.G.G., E.G. and H.A.; investigation, G.G.G., E.G. and H.A.; resources, J.O. and J.M.M.-A.; writing—original draft preparation, G.G.G. and J.M.M.-A.; writing—review and editing, A.L.C.-V., J.O. and J.M.M.-A.; visualization, G.G.G. and J.M.M.-A.; supervision, A.L.C.-V.; project administration, J.O. and J.M.M.-A.; funding acquisition, J.O. and J.M.M.-A. All authors have read and agreed to the published version of the manuscript.

Funding: This work is part of the OptimOre project funded by the European Union Horizon 2020 Research and Innovation Programme under grant agreement No 642201.

Institutional Review Board Statement: Not applicable.

Informed Consent Statement: Not applicable.

Data Availability Statement: Not applicable.

Conflicts of Interest: The authors declare no conflict of interest.

References

1. Fuerstenau, D.W.; Phatak, P.B.; Kapur, P.C.; Abouzeid, A.-Z.M. Simulation of the grinding of coarse/fine (heterogeneous) systems in a ball mill. *Int. J. Miner. Process.* **2011**, *99*, 32–38. [CrossRef]
2. Schönert, K. Aspects of the physics of breakage relevant to comminution. In Proceedings of the 4th Tewksbury Symposium Fracture, February, Melbourne, Australia, 12–14 February 1979; University of Melbourne: Melbourne, Australia, 1979; p. 3.
3. Austin, L.G.; Klimpel, R.R. The Theory of Grinding Operations. *Ind. Eng. Chem.* **1964**, *56*, 18–29. [CrossRef]
4. Deniz, V.; Ozdag, H. A new approach to bond grindability and work index: Dynamic elastic parameters. *Miner. Eng.* **2003**, *16*, 211–217. [CrossRef]

5. Fuerstenau, D.W.; Abouzeid, A.-Z.M. Effect of fine particles on the kinetics and energetics of grinding coarse particles. *Int. J. Miner. Process.* **1991**, *31*, 151–162. [CrossRef]
6. Menéndez-Aguado, J.M.; Coello-Velázquez, A.L.; Dzioba, B.R.; Diaz, M.A.R. Process models for simulation of Bond tests. Transactions of the Institutions of Mining and Metallurgy, Section C. *Miner. Process. Extr. Met.* **2006**, *115*, 85–90. [CrossRef]
7. Coello Velázquez, A.L.; Menéndez-Aguado, J.M.; Brown, R.L. Grindability of lateritic nickel ores in Cuba. *Powder Technol.* **2008**, *182*, 113–115. [CrossRef]
8. Nikolić, V.; Trumić, M. A new approach to the calculation of Bond work index for finer samples. *Miner. Eng.* **2021**, *165*, 106858. [CrossRef]
9. Rittinger von, P.R. *Lehrbuch der Aufbereitungskunde*; Ernst und Korn: Berlin, Germany, 1867.
10. Kick, F. *Das Gesetz der Proportionalen Widerstände und Seine Anwendungen*; Arthur Felix: Leipzig, Germany, 1885.
11. Bond, F.C.; Maxson, W.L. Standard grindability tests and calculations. *Trans. AIME Min. Eng.* **1943**, *153*, 362–372.
12. Bond, F.C. Third theory of comminution. *Trans. AIME Min. Eng.* **1952**, *193*, 484–494.
13. Bond, F.C. Crushing and grinding calculations. *Br. Chem. Eng.* **1961**, *6*, 378–385.
14. Hukki, R.T. Proposal for a solomonic settlement between the theories of von Rittinger, Kick, and Bond. *Trans. AIME* **1961**, *220*, 403–408.
15. Deister, R.J. How to determine the Bond Work Index using lab ball mill grindability tests. *Eng. Min. J.* **1987**, *188*, 42–45.
16. Bearman, R.A.; Briggs, C.A.; Kojovic, T. The Application of Rock Mechanics Parameters to the Prediction of Comminution Behaviour. *Miner. Eng.* **1997**, *10*, 255–264. [CrossRef]
17. Ram Chandar, K.; Deo, S.N.; Baliga, A.J. Prediction of Bond's work index from field measurable rock properties. *Int. J. Miner. Process.* **2016**, *157*, 134–144. [CrossRef]
18. Aras, A.; Özsen, H.; Dursun, A.E. Using Artificial Neural Networks for the Prediction of Bond Work Index from Rock Mechanics Properties. *Miner. Process. Extr. Metall. Rev.* **2020**, *41*, 145–152. [CrossRef]
19. Berry, T.F.; Bruce, R.W. A simple method of determining the grindability of ores. *Can. Min. J.* **1966**, *87*, 63–65.
20. Yap, R.F.; Sepulveda, J.L.; Jauregui, R. Determination of the Bond Work Index Using Ordinary Laboratory Batch Ball Mill. In *Design and Installations of Comminution Circuits*; Mular, A.L., Jergensen, G.V., Eds.; AIME: New York, NY, USA, 1982; p. 176.
21. Bonoli, A.; Ciancabilla, F. The Ore Grindability Definition as an Energy Saving Tool in the Mineral Grinding Processes. In Proceedings of the 2nd International Congress "Energy, Environment and Technological Innovation", Rome, Italy, 12–16 October 1992.
22. Chakrabarti, D.M. Simple approach to estimation of the work index. Transactions of the Institutions of Mining and Metallurgy, Section C. *Miner. Process. Extr. Metall.* **2000**, *109*, 83–89. [CrossRef]
23. Menéndez-Aguado, J.M.; Dzioba, B.R.; Coello-Velázquez, A.L. Determination of work index in a common laboratory mill. *Miner. Metall. Process.* **2005**, *22*, 173–176. [CrossRef]
24. Nikolic, V.; Trumic, M.; Menéndez-Aguado, J.M. Determination of Bond work index in non-standard mills. In Proceedings of the XIV International Mineral Processing and Recycling Conference (2021 IMPRO), Belgrade, Serbia, 12–14 May 2021.
25. Napier-Munn, T.J.; Morrell, S.; Morrison, R.D.; Kojovic, T. *Mineral Comminution Circuits: Their Operation and Optimisation University of Queensland*; Julius Kruttschnitt Mineral Research Centre: Indooroopilly, Australia, 1996.
26. Mwanga, A.; Rosenkranz, J.; Lamberg, P. Development and experimental validation of the Geometallurgical Comminution Test (GCT). *Miner. Eng.* **2017**, *108*, 109–114. [CrossRef]
27. Faramarzi, F.; Napier-Munn, T.; Morrison, R.; Kanchibotla, S.S. The extended drop weight testing approach—What it reveals. *Miner. Eng.* **2020**, *157*, 106550. [CrossRef]
28. Karra, V.K. Simulation of the Bond grindability test. *CIM Bull.* **1981**, *74*, 195–199.
29. Lewis, K.A.; Pearl, M.; Tucker, P. Computer Simulation of the Bond Grindability Test". *Miner. Eng.* **1990**, *3*, 199–206. [CrossRef]
30. Aksani, B.; Sönmez, B. Simulation of Bond grindability test by using Cumulative Based Kinetic Model. *Miner. Eng.* **2000**, *13*, 673–677. [CrossRef]
31. Ciribeni, V.; Bertero, R.; Tello, A.; Puerta, M.; Avellá, E.; Paez, M.; Menéndez-Aguado, J.M. Application of the Cumulative Kinetic Model in the Comminution of Critical Metal Ores. *Metals* **2020**, *10*, 925. [CrossRef]
32. Alfonso, P.; Hamid, S.A.; Anticoi, H.; Garcia-Valles, M.; Oliva, J.; Tomasa, O.; López-Moro, F.J.; Bascompta, M.; Llorens, T.; Castro, D.; et al. Liberation characteristics of Ta-Sn ores from Penouta, NW Spain. *Minerals* **2020**, *10*, 509. [CrossRef]
33. Nava, J.V.; Llorens, T.; Menéndez-Aguado, J.M. Kinetics of dry-batch grinding in a laboratory-scale ball mill of Sn-Ta-Nb minerals from the Penouta Mine (Spain). *Metals* **2020**, *10*, 1687. [CrossRef]
34. Hamid, S.A.; Alfonso, P.; Oliva, J.; Anticoi, H.; Guasch, E.; Sampaio, C.H.; Garcia-Vallès, M.; Escobet, T. Modeling the liberation of comminuted scheelite using mineralogical properties. *Minerals* **2019**, *9*, 536. [CrossRef]
35. Mateus, A.; Figueiras, J.; Martins, I.; Rodrigues, P.C.; Pinto, F. Relative abundance and compositional variation of silicates, oxides and phosphates in the W-Sn-rich lodes of the Panasqueira mine (Portugal): Implications for the ore-forming process. *Minerals* **2020**, *10*, 551. [CrossRef]
36. Alfonso, P.; Tomasa, O.; Garcia-Valles, M.; Tarragó, M.; Martínez, S.; Esteves, H. Potential of tungsten tailings as glass raw materials. *Mater. Lett.* **2018**, *228*, 456–458. [CrossRef]
37. Nava, J.V.; Coello-Velázquez, A.L.; Menéndez-Aguado, J.M. Grinding kinetics study of tungsten ore. *Metals* **2021**, *11*, 71. [CrossRef]

38. GMG -Global Mining Guidelines Group. Determining the Bond Efficiency of Industrial Grinding Circuits. 2016. Available online: https://gmggroup.org/wp-content/uploads/2016/02/Guidelines_Bond-Efficiency-REV-2018.pdf (accessed on 14 May 2021).
39. Braun International Co. *Standard Bond Ball Mill Operating Handbook*; Braun International Co.: Chengdu, China, 1999.
40. Rowland, C.A. Using the Bond work index to measure operating comminution efficiency. *Miner. Metall. Process.* **1998**, *15*, 32–36. [CrossRef]
41. Mosher, J.B.; Tague, C.B. Precision and Repeatability of Bond Grindability Testing. *Miner. Eng.* **2001**, *14*, 1187–1197. [CrossRef]
42. Maxson, W.L.; Cadena, F.; Bond, F.C. Grindability of various ores. *Trans. AIME* **1933**, *112*, 130–145.

Article

Variability of the Ball Mill Bond's Standard Test in a Ta Ore Due to the Lack of Standardization

Gloria González García [1], Alfredo L. Coello-Velázquez [2], Begoña Fernández Pérez [1] and Juan M. Menéndez-Aguado [1,*]

[1] Escuela Politécnica de Mieres, Universidad de Oviedo, Gonzalo Gutiérrez Quirós, 33600 Mieres, Spain; gloria.glez.gcia@gmail.com (G.G.G.); fernandezbegona@uniovi.es (B.F.P.)
[2] CETAM, Universidad de Moa Dr. Antonio Núñez Jiménez, Moa 83300, Cuba; acoello@ismm.edu.cu
* Correspondence: maguado@uniovi.es; Tel.: +34-985-458-033

Abstract: There is no doubt about the practical interest of Fred Bond's methodology in the field of comminution, not only in tumbling mills design and operation but also in mineral raw materials grindability characterization. Increasing energy efficiency in comminution operations globally is considered a significant challenge involving several Sustainable Development Goals (SDGs). In particular, the Bond work index (w_i) is considered a critical parameter at an industrial scale, provided that power consumption in comminution operations accounts for up to 40% of operational costs. Despite this, the variability of w_i when performing the ball mill Bond's standard test is not always understood enough. This study shows the results of a variability analysis (a 3^3 factorial design) performed to elucidate the influence on w_i of several parameters obtained from the particle size distribution (PSD) in feed and product. Results showed a clear variability in the work and grindability indexes with some of the variables considered.

Keywords: comminution; grindability; work index; energy efficiency

1. Introduction

There is no doubt about the importance of Fred Bond's methodology [1–5] and its practical value in the field of comminution, not only in tumbling mills design and operation but also in the characterization of mineral raw materials grindability. The Third Law of Comminution, also known as the Bond's Law, is summarized in Equation (1) [5].

$$W = 10 \cdot w_i \cdot \left(\frac{1}{\sqrt{P_{80}}} - \frac{1}{\sqrt{F_{80}}} \right) \quad (1)$$

wherein:
W is the specific power consumption [kWh/t];
w_i is the Bond work index [kWh/t];
P_{80} is 80% passing size in the grinding product particle size distribution (PSD);
F_{80} is 80% passing size in the feed PSD.

Increasing energy efficiency in comminution operations globally is considered a significant challenge involving several SDGs, especially goals 7 (affordable and clean energy), 9 (industry innovation and infrastructure), 12 (responsible consumption and production) and 13 (climate action), since the increasing energy efficiency reduces waste and emissions production and increases energy availability. In particular, the Bond work index (w_i) is considered a critical parameter at an industrial scale, for power consumption in comminution operations accounts for up to 40% of operational costs [6–8]. Moreover, w_i should be one of the key parameters to consider in a potential process plant digitalization action, using adequate measurable parameters correlation. Despite this, the variability of w_i when performing the ball mill Bond's standard test is not always considered or understood at an industrial scale [9–13]. In the study presented by Mosher and Tague [9], they addressed the variability of Bond test results independent of sampling or procedural variation. They discussed test sensitivity and detailed test procedures to maximize the accuracy and precision of the test, concluding that the Bond tests within one laboratory showed repeatability of less than ±4% at two standard deviations. They also recommended not to report Bond work indices beyond 0.1 kWh/t, based on the precision of the test and suggested that determination of the reproducibility of w_i can be improved significantly by accurate determination of the fresh feed and product PSD. Rodríguez et al. [11] studied this extent, showing that the methodology used for F_{80} and P_{80} determination by interpolation significantly affects w_i calculation.

In the case of the research presented in [10], the results of this research, carried out on a porphyry copper ore, concluded that the Bond work index values differ with different Bond ball mills and with different grinding ball charge distributions, but variations were higher when comparing different Bond ball mills than when comparing different ball charges in the same mill. Maximum variations of 8.6% with different mills and 6.2% with different grinding ball charges were measured.

The authors could not find a precedent comprising a variability study on the Bond standard test itself; mineral processing engineers sometimes attribute the w_i variations to ore grindability changes, while the reason can yield in feed PSD variations. Recently, it has been evidenced that, for a given ore, the grindability function (variation of the Maxon index, gbp, with P_{100}) can present a regular shape while the w_i function with P_{100} can be pretty erratic [14]. Some lack of standardization in the so-called standard test can be the most probable cause of w_i variability. This work presents the result of a careful experimental design defined to elucidate the influence of several parameters obtained from the particle size distribution (PSD) in feed and product on w_i determination.

2. Materials and Methods

2.1. Materials

In order to carry out the series of tests, a 400 kg Ta-Nb-Sn ore sample from the tailings deposit of former mining activities in the Penouta mine (Orense, Spain) was received. A detailed characterization of this ore sample can be found in previous research works [15–17]. The sample was fully sieved in the following size intervals (μm): 3150/2500; 2500/2000; 2000/1600; 1600/1250; 1250/800; 800/500; 500/400; 400/200; 200/160; 160/100. With adequate blending, using the aforementioned size intervals, nine composite feed samples were prepared to fulfil the requirements posed by the multivariate design. In each case, the composite sample was homogenized and divided, checking by PSD analysis that aliquots verified the requirements in each case (Figures S1–S27 at the Supplementary Materials).

2.2. Methods

2.2.1. Bond Ball Mill Standard Test

The procedure to carry out the Bond grindability test [1,18] is described below. The test is performed in the so-called Bond's standard ball mill, a laboratory mill 12″ × 12″, running at 70 rpm (BICO, San Francisco, CA, USA) with rounded inner edges and without lifters. The grinding charge is comprised of a steel balls distribution; Table 1 shows the distribution proposed by Bond in 1961 [5] and that proposed in 1999 [19]; the latter was selected for this test.

Table 1. Evolution of the ball grinding charge distributions proposed by Bond.

Ball Charge Distribution 1961				Ball Charge Distribution 1999			
Ball Size		Balls		Ball Size		Balls	
inch	cm	Number	Weight (g)	inch	cm	Number	Weight (g)
1.45	3.683	43	8803	1.500	3.810	25	5690
1.17	2.972	67	7206	1.25	3.175	39	5137
1.00	2.540	10	672	1.000	2.540	60	4046
0.75	1.905	71	2011	0.875	2.223	68	3072
0.61	1.549	94	1433	0.750	1.905	93	2646
	Total	285	20,125		Total	285	20,592

The mill feed must be prepared by controlled crushing to 100% passing 6 Tyler mesh (3.35 mm). The first grinding cycle feed must be 700 cm^3, and this volume's weight is fixed as the mill charge in all subsequent cycles. Fresh feed PSD is obtained to calculate the 80% passing size (F_{80}) and undersize weight already present in the feed. The test procedure consists of performing several dry grinding cycles to simulate a continuous closed-circuit operation with a 250% circulating load. The circuit is closed by a sieve (P_{100}) selected according to the industrial grinding size target, always between 28 and 325 Tyler mesh (600–45 microns). The detailed grinding cycles procedure can be found in [5,18].

Once finished the grinding cycles, a minimum of five, the ball mill Bond's work index w_i [kWh/sht] can be calculated using Equation (2). In order to express it in metric tons, the corresponding conversion factor must be used.

$$w_i = \frac{44.5}{P_{100}^{0.23} \cdot gbp^{0.82} \cdot \left(\frac{10}{\sqrt{P_{80}}} - \frac{10}{\sqrt{F_{80}}}\right)} \quad (2)$$

where:

w_i is the ball mill Bond's work index [kWh/sht];
P_{100} is the mesh size used to close the grinding circuit [µm];
gbp is the grindability index [g/rev].

It has been recently proposed gbp be renamed as the Maxson index [14]. Walter Maxson led the first research in which gbp was named as the grindability index [1], and was also Fred Bond's mentor at the beginning of his successful career.

2.2.2. Multivariate Experimental Design

The standard test states tight conditions to some test parameters, while others can rest in broad validity ranges. For instance, F_{80} and P_{100} only limitations are being less than 3.35 mm and 600 microns, respectively. Moreover, the undersize content in the ore feed sample is considered by some authors as a variability source. Accordingly, with the same ore, minor differences under correct sampling procedures or even internal procedures in different laboratories could lead to different w_i values. Following the considerations above, the selected variables to perform a variability analysis on the Bond's ball mill standard test were the following:

- Feed particle size, F_{80}
- Closing circuit sieve (should coincide with maximum size in the closed-circuit product, P_{100}
- Undersize percentage in the feed for each P_{100}, % < P_{100}

It is important to notice that F_{80} and the undersize percentage in the feed (% < P_{100}) variations could occur easily due to changes in material preparation; changes in P_{100} should be justified due to changes in the ore liberation size, which is not a strange event in mine operations over time.

Table 2 shows the variables coding (D, C, F) and their values (level 1, 2 or 3) in each case. A total of 27 combinations of variables and levels defined the conditions of the 27 Bond standard tests. Enough ore feed was carefully prepared to fulfil D and F requirements (nine different feed samples prepared), and the Bond standard test was carried out at C value of P_{100} (three levels). It must be understood that, with the same ore and with no further specifications, each of the 27 possibilities fulfils the standard test requirements and the corresponding w_i should be considered with the same validity. The basis and practical use of the ANOVA (SPSS, IBM, Amonk NY, USA) test can be found in Navidi [20].

Table 2. Three levels multivariate experimental design.

Variables		Levels		
		1	2	3
F_{80} (μm)	D	2500	2000	1250
P_{100} (μm)	C	500	400	200
$\% < P_{100}$ (%)	F	0	10	20

3. Results and Discussion

Table 3 collects the results of Bond work index, w_i determination after performing the resulting 27 Bond standard tests; the Mosher and Tague repeatability estimation was considered adequate [9], lower than ±4% at two standard deviations, after checking it with preliminary tests. In Table 3 the *gbp* value obtained in each test is also included. Full details of the performed tests can be found in the spreadsheet file provided as Supplementary Materials.

Table 3. Experimental results of w_i and *gbp*.

	C1								
	D1-F1	D1-F2	D1-F3	D2-F1	D2-F2	D2-F3	D3-F1	D3-F2	D3-F3
w_i [kWh/t]	7.82	8.54	8.96	8.69	9.09	9.50	11.25	11.95	12.13
gbp [g/rev]	6.552	6.008	5.668	6.432	6.265	5.809	6.110	6.046	5.773
	C2								
	D1-F1	D1-F2	D1-F3	D2-F1	D2-F2	D2-F3	D3-F1	D3-F2	D3-F3
w_i [kWh/t]	8.07	8.39	8.45	8.49	8.84	8.80	10.16	10.79	10.69
gbp [g/rev]	5.427	5.220	4.995	5.504	5.383	5.332	5.506	5.377	5.241
	C3								
	D1-F1	D1-F2	D1-F3	D2-F1	D2-F2	D2-F3	D3-F1	D3-F2	D3-F3
w_i [kWh/t]	8.85	9.15	9.29	9.24	9.33	9.46	10.93	11.01	10.50
gbp [g/rev]	3.300	3.157	3.044	3.264	3.235	3.121	3.087	3.082	3.121

The first glance at Table 3 evidences a variability in both w_i and *gbp* values; this variability should be explained due to the sole effect of variables combination in each test. It must be highlighted again that feed preparation was performed carefully, and feed variations among synthetic feeds and a naturally taken feed could be similar to those produced in the field sampling process. In all cases, test conditions fulfilled the Bond standard test requirements (which, in passing, are very open; the only limitation is that feed top size must be under 3.35 mm). Therefore, in summary, the different nine synthetic feeds could be the result of different sampling procedures performed on the same deposit without enough representativity, provided that a tailings pond could show differences in the spatial distribution of particle sizes. Results are also depicted in the Supplementary Materials Figures S28–S30 in the case of w_i, and Figures S31–S33 in the case of *gbp*.

A formal analysis of results was carried out employing the ANOVA test [20], both on w_i and *gbp*. Table 4 garners the ANOVA test results in the case of w_i.

Table 4. Analysis of variance (ANOVA) test results on w_i.

Source of Variance	Sum of Squares	Degrees of Freedom	Mean Square	F-Ratio	p-Value
Main effects					
C	1.9777	2	0.9888	44.99	0
D	30.2763	2	15.1381	688.76	0
F	1.1734	2	0.5867	26.69	0.0003
Interactions					
C&D	2.2241	4	0.5560	25.30	0.0001
C&F	0.5885	4	0.1471	6.69	0.0114
D&F	0.1495	4	0.0374	1.70	0.2422
Residual	0.1758	8	0.0220		
Total (corrected)	36.5654	26			

Table 4 breaks down the variability of w_i into contributions due to individual variables effects and the binary interactions among them. Considering the sum of squares values and p-values in the case of individual variables and binary interactions, variable D (F_{80}) is identified as the primary source of variability among the studied ones. The second source of variability stems from C and D interaction, that is, F_{80} and P_{100} combined effect, which surprisingly has more significant influence than C alone effect. From a w_i variability point of view, F (undersize feed content) was identified as the third variable in importance. In the case of D and F interaction, the p-value is not less than 0.05, so this combination does not have a statistically significant effect on w_i, at the 95.0% confidence level.

Similarly, another ANOVA test was carried out on Maxson grindability index values, and the results are provided in Table 5. In this case, variable C (P_{100}) is identified as the most relevant source of variability; despite D, F and C and F having a p-value more than 0.05 (in consequence, they have a statistically significant effect on gbp, at the 95.0% confidence level), the difference in the sum of squares values lets us affirm that C can be considered as almost the only source of variability in this case.

Table 5. ANOVA test results on gbp.

Source of Variance	Sum of Squares	Degrees of Freedom	Mean Square	F-Ratio	p-Value
Main effects					
C	41.3668	2	20.6834	3653.30	0
D	0.0724	2	0.0362	6.39	0.0220
F	0.5276	2	0.2638	46.59	0
Interactions					
C&D	0.0656	4	0.0164	2.90	0.0937
C&F	0.1921	4	0.0480	8.48	0.0056
D&F	0.0941	4	0.0235	4.15	0.0413
Residual	0.0453	8	0.0057		
Total (corrected)	42.3638	26			

Results suggest that, under the conditions considered in the multivariate design described, the Maxson grindability index, gbp, represents more robustly the intrinsic grindability properties of the ore, being its source of variation the Bond standard test condition, P_{100}. This result reinforces the concept, first proposed by Maxson et al. [1] and subsequently adopted and disseminated by Bond [3–5], that gbp was the best index in characterizing the ore comminution amenability. This fact also justifies the proposal of renaming gbp as the Maxson grindability index.

On the other side, Bond work index variability has a more profound influence from feed PSD conditions (mainly F_{80} value), even to a far greater extent than P_{100} values. As the standard test established relatively frugal recommendations about feed PSD conditions (maximum feed size, F_{100}, less than 3.35 mm), it can be qualified as a worrying source of w_i variation, and the following additional recommendations should be taken into account:

- To establish desirable Bond test conditions, always consider performing feed preparation according to the planned/expected industrial conditions (for instance, by product size estimation on the previous comminution stage—fine crushing or coarse grinding);

- When reporting w_i results, P_{100} and F_{80} values in the test should always be indicated, especially F_{80}, which seems more responsible for w_i variability than P_{100} itself.

4. Conclusions

The following conclusions were derived from this research work and considering the tested ore:

- It was evidenced that the considered parameters induced variability in both Bond work index, w_i, and Maxson grindability index, gbp.
- The ANOVA test results suggested that, in the case of w_i, the primary source of variability is F_{80}, followed by the binary interaction F_{80} and undersize ($<P_{100}$) feed content.
- In the case of gbp, the ANOVA test showed that almost the only source of variability is P_{100}, with almost no influence of feed PSD.
- The following additional recommendations should be taken into account:
- To establish desirable Bond test conditions, always consider performing feed preparation according to the planned/expected industrial conditions
- When reporting w_i results, P_{100} and F_{80} values should always be indicated in the test.

Supplementary Materials: The following are available online at https://www.mdpi.com/article/10.3390/met11101606/s1, Figure S1: Feed PSD, test C1-D1-F1, Figure S2: Feed PSD, test C1-D1-F2, Figure S3: Feed PSD, test C1-D1-F3, Figure S4: Feed PSD, test C1-D2-F1, Figure S5: Feed PSD, test C1-D2-F2, Figure S6: Feed PSD, test C1-D2-F3, Figure S7: Feed PSD, test C1-D3-F1, Figure S8: Feed PSD, test C1-D3-F2, Figure S9: Feed PSD, test C1-D3-F3, Figure S10: Feed PSD, test C2-D1-F1, Figure S11: Feed PSD, test C2-D1-F2, Figure S12: Feed PSD, test C2-D1-F3, Figure S13: Feed PSD, test C2-D2-F1, Figure S14: Feed PSD, test C2-D2-F2, Figure S15: Feed PSD, test C2-D2-F3, Figure S16: Feed PSD, test C2-D3-F1, Figure S17: Feed PSD, test C2-D3-F2, Figure S18: Feed PSD, test C2-D3-F3, Figure S19: Feed PSD, test C3-D1-F1, Figure S20: Feed PSD, test C3-D1-F2, Figure S21: Feed PSD, test C3-D1-F3, Figure S22: Feed PSD, test C3-D2-F1, Figure S23: Feed PSD, test C3-D2-F2, Figure S24: Feed PSD, test C3-D2-F3, Figure S25: Feed PSD, test C3-D3-F1, Figure S26: Feed PSD, test C3-D3-F2, Figure S27: Feed PSD, test C3-D3-F3, Figure S28: Variability of w_i [kWh/t] (P_{100} = 500 µm), Figure S29: Variability of w_i [kWh/t] (P_{100} = 400 µm), Figure S30: Variability of w_i [kWh/t] (P_{100} = 200 µm), Figure S31: Variability of gbp [g/rev] (P_{100} = 500 µm), Figure S32: Variability of gbp [g/rev] (P_{100} = 400 µm), Figure S33: Variability of gbp [g/rev] (P_{100} = 200 µm).

Author Contributions: Conceptualization, J.M.M.-A.; methodology, G.G.G. and J.M.M.-A.; validation, G.G.G. and B.F.P.; formal analysis, G.G.G. and J.M.M.-A.; investigation, G.G.G.; resources, J.M.M.-A.; data curation, A.L.C.-V.; writing—original draft preparation, G.G.G. and J.M.M.-A.; writing—review and editing, B.F.P. and J.M.M.-A.; visualization, G.G.G. and J.M.M.-A.; supervision, A.L.C.-V. and J.M.M.-A.; funding acquisition, G.G.G. and J.M.M.-A. All authors have read and agreed to the published version of the manuscript.

Funding: This work is part of the OPTIMORE project funded by the European Union Horizon 2020 Research and Innovation Programme under grant agreement No. 642201.

Institutional Review Board Statement: Not applicable.

Informed Consent Statement: Not applicable.

Data Availability Statement: Not applicable.

Conflicts of Interest: The authors declare no conflict of interest.

References

1. Maxson, W.L.; Cadena, F.; Bond, F.C. Grindability of various ores. *Trans. AIME* **1933**, *112*, 130–145.
2. Bond, F.C.; Maxson, W.L. Standard grindability tests and calculations. *Trans. AIME Min. Eng.* **1943**, *153*, 362–372.
3. Bond, F.C. Third theory of comminution. *Trans. AIME Min. Eng.* **1952**, *193*, 484–494.
4. Bond, F.C. Crushing and grinding Calculations, Part 1. *Br. Chem. Eng.* **1961**, *6*, 378–385.
5. Bond, F.C. Crushing and Grinding Calculations, Part 2. *Br. Chem. Eng.* **1961**, *6*, 543–548.
6. Rowland, C.A. Using the Bond work index to measure operating comminution efficiency. *Miner. Metall. Process.* **1998**, *15*, 32–36. [CrossRef]

7. Aguado, J.M.M.; Velázquez, A.L.C.; Tijonov, O.N.; Díaz, M.A.R. Implementation of energy sustainability concepts during the comminution process of the Punta Gorda nickel ore plant (Cuba). *Powder Technol.* **2006**, *170*, 153–157. [CrossRef]
8. Coello Velázquez, A.L.; Menéndez-Aguado, J.M.; Brown, R.L. Grindability of lateritic nickel ores in Cuba. *Powder Technol.* **2008**, *182*, 113–115. [CrossRef]
9. Mosher, J.B.; Tague, C.B. Conduct and precision of Bond grindability testing. *Miner. Eng.* **2001**, *14*, 1187–1197. [CrossRef]
10. Kaya, E.; Fletcher, P.C.; Thompson, P. Reproducibility of Bond grindability work index. *Miner. Metall. Process.* **2003**, *20*, 140–142. [CrossRef]
11. Rodríguez, B.Á.; García, G.G.; Coello-Velázquez, A.L.; Menéndez-Aguado, J.M. Product size distribution function influence on interpolation calculations in the Bond ball mill grindability test. *Int. J. Miner. Process.* **2016**, *157*, 16–20. [CrossRef]
12. Makhija, D.; Mukherjee, A.K. Effect of undersize misplacement on product size distribution of Bond's ball mill test. *Miner. Process. Extr. Metall.* **2016**, *125*, 117–124. [CrossRef]
13. Magdalinovic, N.; Trumic, M.; Trumic, G.; Magdalinovic, S.; Trumic, M. Determination of the Bond work index on samples of non-standard size. *Int. J. Miner. Process.* **2012**, *114–117*, 48–50. [CrossRef]
14. García, G.G.; Oliva, J.; Guasch, E.; Anticoi, H.; Coello-Velázquez, A.L.; Menéndez-Aguado, J.M. Variability study of bond work index and grindability index on various critical metal ores. *Metals* **2021**, *11*, 970. [CrossRef]
15. Alfonso, P.; Tomasa, O.; Garcia-Valles, M.; Tarragó, M.; Martínez, S.; Esteves, H. Potential of tungsten tailings as glass raw materials. *Mater. Lett.* **2018**, *228*, 456–458. [CrossRef]
16. Alfonso, P.; Hamid, S.A.; Anticoi, H.; Garcia-Valles, M.; Oliva, J.; Tomasa, O.; López-Moro, F.J.; Bascompta, M.; Llorens, T.; Castro, D.; et al. Liberation characteristics of Ta-Sn ores from Penouta, NW Spain. *Minerals* **2020**, *10*, 509. [CrossRef]
17. Nava, J.V.; Coello-Velázquez, A.L.; Menéndez-Aguado, J.M. Grinding kinetics study of tungsten ore. *Metals* **2021**, *11*, 71. [CrossRef]
18. GMG-Global Mining Guidelines Group. Determining the Bond Efficiency of Industrial Grinding Circuits. 2016. Available online: https://gmggroup.org/wp-content/uploads/2016/02/Guidelines_Bond-Efficiency-REV-2018.pdf (accessed on 29 September 2021).
19. BICO (Braun International Co.). *Standard Bond Ball Mill Operating Handbook*; BICO: San Francisco, CA, USA, 1999.
20. Navidi, W. *Statistics for Engineers and Scientists*, 5th ed.; McGraw-Hill Education: New York, NY, USA, 2019.

Article

A Comparative Study of Energy Efficiency in Tumbling Mills with the Use of Relo Grinding Media

Nikolay Kolev [1], Petar Bodurov [1], Vassil Genchev [1,†], Ben Simpson [2], Manuel G. Melero [3] and Juan M. Menéndez-Aguado [4,*]

[1] Relo-B Ltd., 1463 Sofia, Bulgaria; nicolai.kolev@mail.bg (N.K.); pbodurov@mgu.bg (P.B.); vassil.genchev@digitalprint.bg (V.G.)
[2] Wardell Armstrong International Ltd., Truro TR3 6EH, UK; bsimpson@wardell-armstrong.com
[3] Departamento de Ingeniería Eléctrica, Electrónica, de Computadores y Sistemas, Universidad de Oviedo, 33203 Gijón, Spain; melero@uniovi.es
[4] Escuela Politécnica de Mieres, Universidad de Oviedo, 33600 Mieres, Spain
* Correspondence: maguado@uniovi.es; Tel.: +34-985458033
† In memory of V. Genchev.

Abstract: An evaluation of Relo grinding media (RGM, Reuleaux tetrahedron-shaped bodies) performance versus standard grinding media (balls) was made through a series of grinding tests, including a slight modification of the standard Bond test procedure. Standard Bond tests showed a reduction in the Bond ball mill work index (w_i) of the mineral sample used in this study when using Relo grinding media. The modified Bond test procedure is based on using the standard Bond ball work index test but changing the circulating loads (350%, 250%, 150%, 100%). The comparative tests with RGM were carried out at the same number of revolutions as the grinding tests with balls at respective circulating load. The RGM charge yielded a 14% higher net undersize product than balls, which hints at improving energy efficiency and the potential for significant mining industry benefits.

Keywords: comminution; ball mills; grinding media; energy efficiency

1. Introduction

Energy efficiency in the mining industry is an ever-growing concern for sustainable mineral processing and the life of mine activities. Prior to beneficiation, the ore must pass through various stages of comminution, the process in which the particle size of the ore is progressively reduced until mineral particles have been liberated. Comminution operations, including grinding, consume up to 4% of electrical energy globally, and about 50% of mine site energy consumption is in comminution [1–6]. A study shows that the grinding process alone contributes to approximately 40% of all power consumption in a mine complex [7]. Tumbling mills are notorious for their low energy efficiency because they only use up to 10% of installed power for grinding action. A feature of ball mills is their high specific energy consumption; a mill filled with balls, working idle, consumes approximately as much energy as at full-scale capacity, i.e., during the grinding of material. Radziszewski [8] showed that 56% of the input energy in grinding circuits becomes heat lost to the environment; 43% is lost in heating the slurry, while only 1% is actual breakage energy. A recent and more comprehensive study has shown that, on average, 79% of the supplied electrical energy converts to heat absorbed by the slurry, 8% is lost through the drive system; about 2% of the energy is transmitted to ambient air, and just about 10% is used for the grinding work [9]. Therefore, about 90% of the thermal energy is potentially recoverable, and there is vast potential for improvement in energy efficiency.

One way to reduce energy and material consumption in milling is to design and select the grinding media properly. Hassanzadeh [10] argues that ball size distribution plays a significant role in energy consumption and ball mill efficiency. Larger balls break coarse particles mainly by impact, while smaller balls produce breakage by abrasion. Generally, for a fixed volume of grinding charge, particle–ball collision frequency falls rapidly as the ball size increases. Research about mixtures of media shapes points out that, by combining different grinding mechanisms in terms of contacts, the volume of grinding zones can be efficiently increased when there is an optimal mixture of two or more grinding media with different shapes and, therefore, the milling kinetics can be improved [11].

There have been several attempts to establish the shape of grinding media that is best suited for tumbling mills. One of the most comprehensive studies [12] concluded that the sphere is the most efficient for all shapes tested for constant batch grinding time. Kelsall et al. [13] studied the influence of different grinding media shapes (steel spheres, cubes, equi-cylinders, and hexagonal "cylinders"), and they also concluded that spherical media handled the most significant throughput and produced the most correctly sized product. Many authors compared spherical balls to cylpebs, and they concluded that balls had better grinding efficiency [14–17]. Some authors claimed that cylpebs had produced slightly more fines than balls, but no quantification of increased throughput has been made [18,19]. In [20], the grinding behavior of balls, eclipsoids, and cubes was investigated. In the case of eclipsoids, which have a 25% higher total surface area than balls equivalent in volume, it was found that the increase in surface area available for breakage does not necessarily translate into an increase in the breakage rate. This study concluded that balls proved to have a higher rate of breakage, confirming that balls are the most efficient grinding media.

Although there are no quantified statements on what range of improvement in energy efficiency has been achieved for tumbling mills when changing the shape of grinding media in tumbling mills, cylpeb manufacturers claim that they deliver 25% greater grinding capacity in a typical mill charge. However, this may apply only to regrind milling circuits. Relo grinding media (RGM) [21], which are described in Section 2, seem to be suitable for a wide range of applications, including SAG mills. Moreover, in [21], a plant trial is described in which RGM increased mill throughput by 80%, achieving approximately 45% lower energy consumption than balls.

This research work aims to test the performance of RGM in a laboratory mill, quantifying the differences when using RGM instead of balls.

2. Physical Properties of Relo Grinding Media

The introduced RGM (Figure 1) receives the name in honor of the German engineer Franz Reuleaux, a mechanical engineer, who gave his name to this geometrical shape in the nineteenth century. Although these grinding bodies come in slightly different shapes, they are all derived from Reuleaux geometry, i.e., the Reuleaux triangle and the Reuleaux tetrahedron are the base structural shapes of RGM. RGM are made of steel, including all types commonly used for making balls. Moreover, according to Penchev and Bodurov [22], due to their shape, RGM steel bodies have better quench hardening than balls at equal tempering conditions; these comparative tests were performed using equal steel (0.65% C, 1.03% Mn). Accordingly, it seems appropriate to study the performance of RGM and see if they can offer significant efficiency improvements in tumbling mill grinding circuits.

Because of their Reuleaux geometry, an RGM charge has a greater surface area and a higher bulk density than balls with the same mass and size. Table 1 shows the comparative physical properties for RGM and balls. Ideally, RGM have a 10% greater surface than balls of the same mass (volume) and 10% higher bulk density than steel balls. As a result, for a given charge volume, more grinding media surface area is available for size reduction when charged with RGM, but the mill would also draw more power.

(a) (b)

Figure 1. Grinding media charges: (**a**) Relo M-1 vs. balls (**b**) Relo M-2 vs. ball.

Table 1. Comparative data of physical properties for RGM and balls.

RGM (γ = 7.85 g/cm^3)					Balls (γ = 7.85 g/cm^3)				
Edge Arc (mm)	Mass (g)	Surface Area (cm^2)	Specific Surface (cm^2/g)	Bulk Density (t/m^3)	Diameter (mm)	Mass (g)	Surface Area (cm^2)	Specific Surface (cm^2/g)	Bulk Density (t/m^3)
41.4	235	51	0.22	5.2					
40.2	212	48.1	0.23	5.2	38.1	224	45.6	0.20	4.6
33.5	112	33.4	0.28	5.2	31.8	128	31.8	0.25	4.6
26.7	61.5	21.1	0.35	5.2	25.4	65.5	20	0.31	4.6
23.8	43	16.9	0.39	5.2	22.2	43.5	15.5	0.36	4.6
20.1	25.8	12.1	0.47	5.2	19.1	27.3	11.5	0.42	4.6

RGM sizes were intentionally selected for the Bond mill tests such that they have both lower mass and greater surface area. The research objective was to show that surface area is the main factor for the higher grinding efficiency of RGM instead of the total mass. Table 2 shows selected mill charges with the same number of grinding elements.

Table 2. Bond mill test media charge—5% greater surface area.

Ball Diameter/Relo Edge Arc, [mm]	38.10/40.2	31.8/33.4	25.4/26.7	22.2/23.8	19.05/20.1	Total Number	Total Mass, g
Ball Charge—number of balls	25	39	62	69	90	285	20,115
RGM Charge—number of relos	25	39	62	69	90	285	19,149

3. Materials and Methods

This research's main goal was to collect and analyze data from comparative laboratory grinding tests to compare the grinding performance of RGM with a conventional ball charge in a Bond ball mill (BICO BRAUN, Burbank, CA, USA). The tests are discussed in more detail below.

3.1. Media Charge Conditions

Three media conditions were considered to have comparable and repeatable results:
- Media mass,
- Media surface area, and
- Media size distributions.

For the comparative tests, the standard Bond ball charge was set as the base case. The RGM charge was controlled to match two of the three media conditions of the balls, with the third one being different (surface area), to distinguish one effect in each test. Table 2 presents the details of both media charge size distributions.

Ball charge (Table 2) is the standard Bond ball mill charge condition, which was used as the base case for comparison. RGM charge (Table 2) was defined to ensure that the RGM have a similar size but significantly smaller mass. Table 3 presents an RGM charge in which six smaller Relo grinding bodies (26.7 mm) were omitted, and six larger grinding Relo bodies were added to give equal mass to the balls. This RGM charge has a similar mean size and mass in each size fraction (hence the total mass), but due to their shape, there is approximately 10% more surface area than that of balls.

Table 3. Bond mill media charge—equal total mass, 10% greater surface area.

Ball Diameter/Relo Edge Arc [mm]	40/41.4	38.10/40.2	31.8/33.4	25.4/26.7	22.2/23.8	19.05/20.1	Total Number	Total Mass [g]
Number of balls	0	25	39	62	69	90	285	20,115
Number of RGM	6	25	39	56	69	90	285	20,165

3.2. Ore Sample

The sample selected was a pegmatite, which is a spodumene ore. XRF results are shown in Table 4, and the feed particle size distribution (PSD) is shown in Table 5. It was composed primarily of quartz and plagioclase, with minor amounts of lepidolite, beryl, and potassium feldspar. The ore sample was rather hard rock, with w_i = 16.8 kWh/t, and was prepared using a jaw crusher and a rotary divider to obtain the representative subsamples for each test.

Table 4. Feed chemical analysis (L.O.I. = lost on ignition).

Element	SiO_2	Al_2O_3	Na_2O	Li_2O	K_2O	Fe_2O_3	CaO	L.O.I.
(%)	72.16	17.68	6.56	1.20	0.92	0.50	0.09	0.89

Table 5. Particle-size distribution analysis.

Size (μm)	3350	2000	1500	1000	800	600	400	300	200	150	100	75
Cum. Passing (%)	100.00	98.69	88.42	66.93	49.66	38.19	28.12	20.77	14.37	9.44	4.66	1.14

3.3. Power Draw Measurement

To compare the power draw required by the Bond mill in the grinding process with RGM versus balls, the active power consumption (watts) of the electric motor driving the mill was measured. This assumption is based on the fact that no significant differences can be found between both grinding media for the electric and mechanical losses of the motor, and, consequently, the measurement of the active power can be used to compare the mechanical power draw.

The electric driving machine is a single-phase motor, so to evaluate its electric consumption, a measurement of the supply voltage and current is needed. Then, the active power is obtained according to:

$$P = \frac{1}{T} \int_0^T v(t)i(t)dt \tag{1}$$

with P the active power, T the period, $v(t)$ the instantaneous value of the supply voltage, and $i(t)$ the instantaneous value of the supply current. A differential probe with ×500 attenuation was used for voltage signal measuring. In the case of the current signal, a probe providing a transformation of 100 mV/A was used. Both signals were recorded and processed employing a 60 MHz bandwidth digital scope.

3.4. Test Procedure

The Bond work index, w_i, is in everyday use for the assessment of comminution efficiency, and it has been generally accepted as a measure of material grindability (ores, cement clinker). In this research, the Bond test procedure was considered fundamental for comparing the milling performance of different grinding media shapes. However, as some authors noted [22], there is a problem in designing a locked-cycle test procedure for such a purpose since the Bond test does not take the actual energy usage into account. The standard Bond test procedure [2,23] requires stopping the test when the grindability reaches equilibrium, which is typically achieved at 3% variation between the last two runs. Since the RGM charge may require fewer mill revolutions than the ball charge, this indicates that the grinding time in the two tests would be different. Thus, the energy input for the ball charge test would be greater than for the RGM charge test.

With the aim of predicting the grinding performance of RGM in a full-scale ball mill, a simplified procedure for the scale-up of a ball mill was adopted. This procedure involved laboratory tests using the Bond ball mill and test conditions to simulate the full-scale mill circuit's steady-state performance from laboratory results. Accordingly, all the locked-cycle laboratory tests were conducted in a standard Bond ball mill loaded with two types of steel grinding media—balls and RGM—to treat the same feed ore at a time.

Two types of comparative locked-cycle tests were conducted:

- Test series 1: Standard Bond work index tests using the standard set of balls and RGM charge (Table 2).
- Test series 2: Comparative grinding tests using balls and RGM at equal mass and media size distribution (Table 3) but at different circulating loads.

The test procedure for the first series of grinding tests follows the well-known Bond ball mill grindability test [2,23] precisely. In the Bond ball milling test, a locked-cycle test, the fresh feed to the test is crushed down to 100% under 3.35 mm. The mill grinds a constant 700 mL of ore. After each grind, the mill contents are screened to remove undersize and replenished with an equal mass of new feed. The length of grinding time for each run is adjusted until the oversize fraction's mass is consistently 2.5 times greater than the undersize. Under these conditions, the test approximates a closed-circuit continuous mill's steady-state performance with a circulating load of 250%.

The second series of locked-cycle grinding tests were run at four different circulating loads of 100%, 150%, 250%, and 350%. Since the RGM charge required fewer mill revolutions than the ball charge [24], a slight change in the methodology was needed to get a better comparative analysis. The modification consisted of adding more fresh feed to the tests using RGM to get the same circulating load (250%) at the same number of mill revolutions of both ball charge and RGM charge tests. To compensate for the lower circulating load when RGM was used, the amount of fresh sample was increased at the same number of revolutions as per the standard tests with balls. The test conditions for both locked-cycle tests are summarized in Table 5. Since the grindability index (net grams of screen undersized product per mill revolution, Gpr) is the primary variable to determine the Bond work index [2], it was used as a comparative measurement to show the difference between RGM charge and ball charge milling performance.

4. Results and Discussion

4.1. Power Draw Measurements

Since the bulk density of an RGM charge was approximately 10% higher than the bulk density of a ball charge, the two media charges' torques should be different. Therefore, it was logical to expect that there must be a difference in the power draw of these two types of grinding media. According to Lameck's results [17], cylpebs draw approximately 30%

less power than balls at 90% of critical speed (Figure 2), so it was interesting to carry out a similar comparative study between RGM and balls.

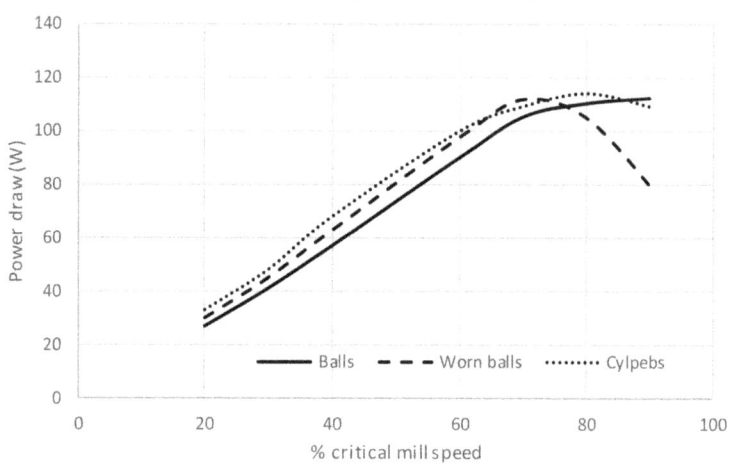

Figure 2. Power variation with mill speed for different media shapes (adapted from [16]).

Measurements performed (see Table 6) confirmed that RGM bodies draw higher power than balls without mineral charge, similar power as balls at standard conditions of a Bond test (35% of voids within the ball charge filled with mineral charge), and less power than balls. It is important to note that RGM and balls may draw the same power at a given % of critical speed, but grindability might not be equal. These grinding tests were performed for 2 min, the power draw deviation in all cases being below 1.5%.

Table 6. Power draw measurements in a Bond ball mill.

Bond Mill Charge	Power Draw without Ore Sample [W]	Power Draw at 35% of Voids within the Ball Charge [W]	Power Draw, at 100% of Voids within the Ball Charge [W]
No load	309	-	-
Standard ball charge (20.1 kg)	402	437	462
RGM charge (20.1 kg)	424	440	445

Considering that the Bond methodology uses a filling ratio in the standard test ball mill (35%), which is relatively lower than the filling ratio at industrial scale (where the ideal situation is when the mineral charge fills 100% of voids), and considering that the influence of mineral filling ratio is clearly stronger in the case of balls than in the case of RGM charge, it could be expected that the Bond methodology would overestimate the specific energy consumption in the case of using RGM charge. Further research should be performed to define a coefficient that lets an RGM work index be obtained from standard Bond tests.

4.2. Test Series 1

The main goal of test series 1 was to see whether the lower mass (5% lower) of the RGM charge would affect the grinding performance and calculation of w_i. The RGM charge was designed to have a 5% lower total weight but a 5% higher total surface area. In both cases, w_i values were calculated using the formula proposed by Bond in the standard test (Equation (2)).

$$w_i [kWh/sht] = \frac{44.5}{P_{100}^{0.23} \cdot Gpr^{0.82} \cdot \left(\frac{10}{\sqrt{P_{80}}} - \frac{10}{\sqrt{F_{80}}}\right)} \qquad (2)$$

Tests results are summarized in Table 7. Standard tests have shown that the RGM charge achieves a 2% lower w_i than balls at equal testing conditions (number of revolutions, mill speed, ore sample, closing screen aperture). This result cannot be considered conclusive, for it is widely accepted [5] that the repeatability error of the standard Bond test is around 3.5%. However, it must be taken into account that the RGM charge was 5% lighter than the ball charge when applying Bond's methodology to RGM.

Table 7. Standard Bond grindability tests parameters and results (Test series 1).

Test Parameter	Unit	Balls	RGM
Weight of media charge	(kg)	20.123	19.145
P_{100}	(microns)	106	106
Gpr	(g/rev)	1.201	1.190
F_{80}	(microns)	2044	2044
P_{80}	(microns)	84	81
Result	**Unit**	**Balls**	**RGM**
w_i	(kWh/t)	16.62	16.30

Moreover, if a similar work index is obtained using a 5% lighter steel charge, this means that less grinding charge weight can produce similar grinding work, so a better energy conversion is produced when using RGM charges.

4.3. Test Series 2

It is widely accepted that the higher the circulating load of a grinding circuit is, the lower the probability of ultrafine particle production. It means that the most efficient ball milling circuits require a high circulating load ratio (CLR) [16]. By maintaining a high percentage of coarse solids in the mill, a high circulating load results in a much more efficient grinding circuit. The purpose of our grinding tests at various circulating loads was to see how this could apply to new grinding media compared to standard grinding media. These tests clearly show the advantages of RGM over balls at different circulating loads (Figure 3).

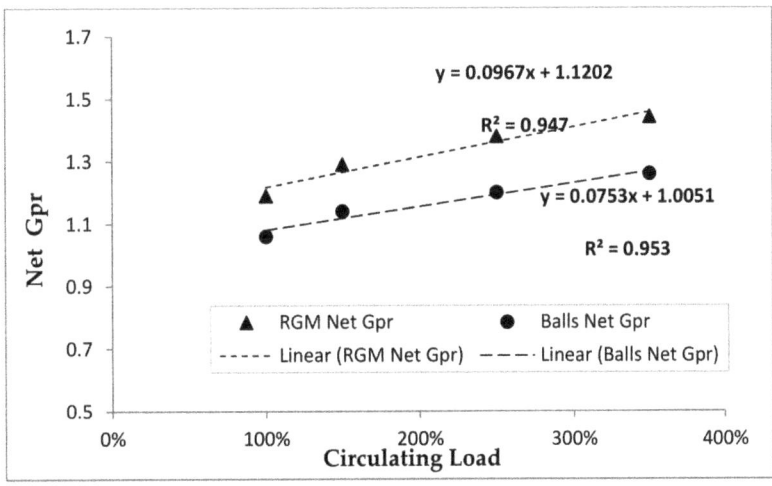

Figure 3. Grinding tests at different CLR—Net Gpr vs. CLR.

The results show that, using RGM, the most efficient ball milling circuits may not necessarily require a high circulating load ratio. The data revealed that the RGM at

100% circulating load and balls at 250% circulating load have almost equal productivity (grinding rate), which means that the RGM charge is more energy-efficient and will require considerably less power for classification (cyclone feed pumping). McIvor [25,26] wrote that 250% circulating load pumping accounts for about 8% of total grinding costs. Therefore, the replacement of balls with RGM will reduce total grinding costs at the same level of mill productivity.

On the other hand, if energy consumption is kept equal, the RGM charge achieves on average 14% higher productivity (measured in Gpr) than balls at the corresponding circulating load.

As previously discussed, these four pairs of tests at different CLRs were performed in a Bond mill. The mass of two mill charges and the number of revolutions per given circulating load was equal, and mill speed was kept constant (the one in the standard test mill, 70 rpm). More feed was added to the Relo tests to reach the steady state (desired circulating load) due to the Relo media's apparent higher grinding kinetics.

The results from these tests were studied using linear regression equations (Figures 3 and 4). In Figure 3, the circulating load ratio (%) is represented in abscises while net Gpr is represented in ordinates. This linear regression model shows that, if circulating load is increased by 1%, then the net Gpr of a mill loaded with RGM will be expected to increase by 0.097 g/rev, and if there were no circulating load, we would expect a net Gpr of 1.12 g/rev. For balls, net Gpr will be expected to increase by 0.075 g/rev if the circulating load is increased by 1%; with no circulating load, the ball mill's net Gpr should be 1 g/rev. Moreover, comparing the situation at high circulating loads (350%), RGM net Gpr is 26% higher than ball Net Gpr, which means that, under similar conditions, a mill throughput should increase by that percentage when grinding with a RGM charge. This difference can be even greater at industrial scale, since for the same fraction of the critical speed, a full-scale mill's rotational speed is lower than that of a pilot-scale mill [24]. Thus, the breakage rate of a full-scale mill is higher at the smaller sizes because there are more grinding media layers present in the bulk of the charge in full-scale mills [24]. McIvor [25,26] pointed out that the material's size distribution going through the ball mill was much coarser at a high CLR than at a low CLR.

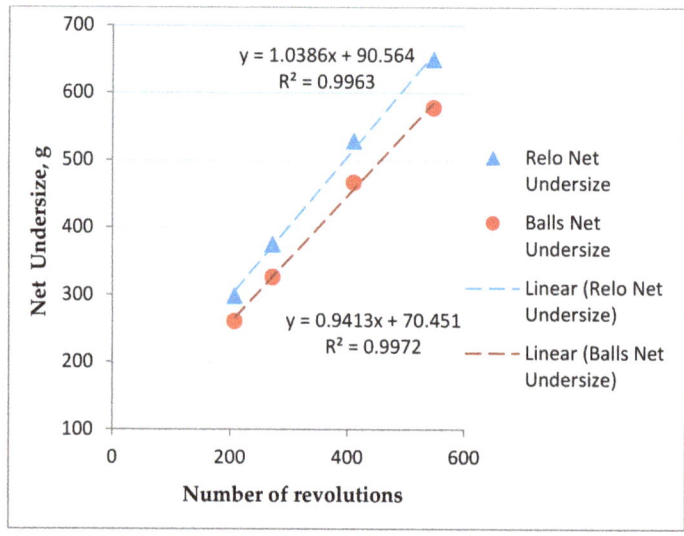

Figure 4. Grinding tests at different CLR—net undersize vs. number of revolutions.

In Figure 4, the number of revolutions versus the net undersize is plotted. The linear regression model shows that, at a large number of mill revolutions (longer grinding cycles, higher residence time), the net undersize of Relo grinding media will be approximately 10.5% higher than the net undersize of balls. So when the circulating load is zero, the RGM charge productivity is 10–12% higher.

The higher fine-particle production with the RGM charge corresponds perfectly with the contact spots ratio between two Relo bodies and two spheres. According to Penchev [22], a 40 mm radius sphere would have the same weight as a Reuleaux spheroidal tetrahedron with spherical wall 87 mm in radius; and the contact area between two Relo grinding bodies will be 29% larger than between two balls. Hence, the probability of collisions of RGM with an ore particle will be 29% higher than in balls.

4.4. Optimum Economic Circulating Ratio (Trade-Off between RGM and Balls at Circulating Loads)

Increasing pumping energy (pump and cyclone maintenance costs included) should be balanced against decreasing grinding energy and media costs when circulating load increases, so the optimum economic circulating load can be identified (Table 8).

Table 8. Trade-off between ball charge and RGM charge in terms of circulating load.

Media	Mill Throughput [tph]	CLR [%]	Relative Grinding Costs [%]	Relative Pumping/Classification Costs [%]	Total [%]
	Base case				
Balls	100	245	92	8	100
Relo	100	100	92	4	96
	Increased by 14%				
Balls	114	558	84	16	100
Relo	114	245	84	8	92

If the RGM mill charge and the ball mill charge draw the same power of the mill motor and have equal mill throughput (100 tph), the grinding circuit operating costs with RGM charge should be 4% lower (Table 8), mainly due to the lower circulating load of the RGM circuit. However, under a 14% production increase scenario, the CLR should increase from 245% to 558%, thus increasing pumping/classification costs and increasing total operating costs up to 8%.

4.5. Correction Coefficients for Bond WI Using RGM

Finally, looking at the interpretation of these laboratory test results, we may conclude that RGM behavior and milling performance are very different in an industrial mill than in a laboratory scale [21]. Plant test results show drastically lower energy consumption (kWh/t) of RGM than laboratory test data suggest. The explanation can be found by looking at the difference between breakage rate distributions obtained from the pilot-scale tests and full-scale mills [27]. Yu [28] wrote that, for the same fraction of the critical speed, a full-scale mill's rotational speed is lower than that of a laboratory-scale mill. Thus, at the coarse sizes, the breakage rates in a laboratory-scale mill are higher than that in a full-scale mill. A full-scale mill's breakage rate is higher at the smaller sizes because there are more grinding media layers present in the bulk of the charge in full-scale mills. These laboratory grinding tests with RGM may not be directly used to predict a RGM-loaded full-scale mill.

A new model is needed to describe an industrial tumbling mill charged with RGM, and it is required to introduce a correction factor to apply Bond's methodology. It is important to consider that a reduction in the Bond work index, maintaining the rest of the conditions, would reduce the circulating load, increase the fresh feed throughput to the closed grinding circuit and eventually increase grinding production. Based on this test program, we can conclude that higher net undersize will reduce w_i by a certain percentage. The lower power draw of RGM charge and reduced w_i by RGM mean that the total reduction of the standard Bond w_i will need to be justified by correction coefficients.

Using this interdependence, we show that RGM will yield higher throughput than balls at similar grind size at the same power draw. The advantage of the higher efficiency could be achieved in different ways:

- By maintaining current throughput with a smaller grinding size (i.e., when the liberation size decreases);
- By increasing throughput, maintaining the grinding size;
- By reducing the filling ratio of the ball mill, thus lowering power and grinding media consumption.

5. Conclusions

- Power measurement tests evidenced differences in RGM and ball energy performance. Further research should be carried out to define a coefficient to obtain an RGM work index from standard Bond tests.
- Standard Bond tests did not show a clear improvement in the energy efficiency of the RGM charge compared to balls. The Bond ball work index using the RGM charge was 2% lower, while the repeatability error for the standard Bond test is estimated to be below 3.5%.
- Grinding tests at various CLRs and the same grinding time at each other circulating load test revealed that the grinding rate of the RGM charge at 100% circulating load is the same as the grinding rate of balls at 250% of circulating load.
- Linear regression calculations suggested that, at a low number of mill revolutions (equal to high circulating load conditions), RGM need 50% less grinding time than balls to produce the same amount of undersize. It showed that, working at coarser feed (high circulating loads), RGM could be more efficient than balls, thus lowering the power consumption of tumbling mills. The mass of undersize products from these tests was 14% higher on average when the RGM charge was used.
- The trade-off carried out between RGM and balls at circulating loads showed a significant improvement in energy efficiency if using RGM when facing a throughput increase, mainly due to the reduction in operating costs.

Author Contributions: Conceptualization, N.K. and J.M.M.-A.; methodology, N.K., B.S. and J.M.M.-A.; validation, P.B. and V.G.; formal analysis, N.K. and J.M.M.-A.; investigation, B.S. and M.G.M.; resources, N.K., B.S. and J.M.M.-A.; data curation, N.K.; writing—original draft preparation, N.K.; writing—review and editing, B.S., J.M.M.-A. and M.G.M.; visualization, N.K., P.B. and V.G.; supervision, P.B. and V.G.; funding acquisition, N.K. and M.G.M. All authors have read and agreed to the published version of the manuscript.

Funding: This research was partially funded by the Spanish Ministry of Economy and Competitiveness, under project DPI2017-83804-R.

Institutional Review Board Statement: Not applicable.

Informed Consent Statement: Not applicable.

Data Availability Statement: Not applicable.

Acknowledgments: Wardell Armstrong International Ltd. is gratefully acknowledged for performing the required laboratory test work. Authors also thank Donald Leroux and Ahmad Hassanzadeh for their comments and advice regarding RGM performance.

Conflicts of Interest: The authors declare no conflict of interest.

References

1. Jankovic, A.; Valery, W.; La Rosa, D. Fine Grinding in the Australian Mining Industry. In Proceedings of the 3rd International Conference on Recent Advances in Materials, Minerals and Environment (RAMM 2003), Nibong Tebal, Malaysia, January 2003; Universiti Sains Malaysia: Penang, Malaysia.
2. Bond, F.C. Crushing and Grinding Calculations Parts 1 and 2. *Br. Chem. Eng.* **1961**, *6*, 378–385, 543–548.

3. Pedrayes, F.; Norniella, J.G.; Melero, M.G.; Menéndez-Aguado, J.M.; del Coz-Díaz, J.J. Frequency domain characterization of torque in tumbling ball mills using DEM modelling: Application to filling level monitoring. *Powder Technol.* **2018**, *323*, 433–444. [CrossRef]
4. Osorio, A.M.; Menéndez-Aguado, J.M.; Bustamante, O.; Restrepo, G.M. Fine grinding size distribution analysis using the Swebrec function. *Powder Technol.* **2014**, *258*, 206–208. [CrossRef]
5. Rodríguez, B.Á.; García, G.G.; Coello-Velázquez, A.L.; Menéndez-Aguado, J.M. Product size distribution function influence on interpolation calculations in the Bond ball mill grindability test. *Int. J. Miner. Process.* **2016**, *157*, 16–20. [CrossRef]
6. Ciribeni, V.; Bertero, R.; Tello, A.; Puerta, M.; Avellá, E.; Paez, M.; Menéndez Aguado, J.M. Application of the Cumulative Kinetic Model in the Comminution of Critical Metal Ores. *Metals* **2020**, *10*, 925. [CrossRef]
7. Ballantyne, G.R.; Powell, M.S.; Tiang, M. Proportion of Energy Attributable to Comminution. In Proceedings of the 11th AusIMM Mill Operator's Conference, Hobart, Australia, 29–31 October 2012.
8. Radziszewski, P. Energy recovery potential in comminution processes. *Miner. Eng.* **2013**, *46–47*, 83–88. [CrossRef]
9. Bouchard, J.; LeBlanc, G.; Levesque, M.; Radziszewski, P.; Georges-Filteau, D. Breaking Down Energy Consumption in Industry Grinding Mills. In *Proceedings of the 49th Annual Canadian Minerals Processors Conference, Ottawa, ON, Canada, 17–19 January 2017*; Muinonen, J., Cameron, R., Zinck, J., Eds.; Canadian Institute of Mining, Metallurgy and Petroleum (CIM): Westmount, QC, Canada, 2017; pp. 25–35.
10. Hassanzadeh, A. The Effect of Make-Up Ball Size Regime on Grinding Efficiency of Full-Scale Ball Mill. In Proceedings of the XVII Balkan Mineral Processing Congress, Antaliya, Turkey, 1–3 November 2017; Volume 1, pp. 117–124.
11. Simba, K.P.; Moys, M. Effects of mixtures of grinding media of different shapes on milling kinetics. *Miner. Eng.* **2014**, *61*, 40–46. [CrossRef]
12. Norris, G.C. Some grinding tests with spheres and other shapes. *Trans. Inst. Miner. Metall.* **1954**, *63*, 197–209.
13. Kelsall, D.F.; Stewart, P.S.B.; Weller, K.R. Continuous grinding in a small wet ball mill Part 5. A study of the influence of media shape. *Powder Technol.* **1973**, *8*, 77–83. [CrossRef]
14. Cloos, U. Cylpebs: An alternative to balls as grinding media. *World Min.* **1983**, *36*, 59.
15. Herbst, J.A.; Lo, Y.C. Grinding efficiency with balls or cones as media. *Int. J. Miner. Process.* **1989**, *26*, 141–151. [CrossRef]
16. Shi, F. Comparison of grinding media-Cylpebs versus balls. *Miner. Eng.* **2004**, *17*, 1259–1268. [CrossRef]
17. Lameck, N.N.S. Effects of Grinding Media Shapes on Ball Mill Performance. Master's Thesis, Faculty of Engineering and The Built Environment, University of the Witwatersrand, Johannesburg, South Africa, 2006.
18. Ipek, H. Effects of grinding media shapes on breakage parameters. *Part. Part. Syst. Charact.* **2007**, *24*, 229–235. [CrossRef]
19. Cuhadaroglu, D.; Samanli, S.; Kizgut, S. The effect of grinding media shape on the specific rate of breakage. *Part. Part. Syst. Charact.* **2008**, *25*, 465–473. [CrossRef]
20. Simba, K.P. Effects of Mixture of Grinding Media of Different Shapes on Milling Kinetics. Ph.D. Thesis, Faculty of Engineering and the Built Environment, University of the Witwatersrand, Johannesburg, South Africa, 2010. Available online: https://core.ac.uk/download/pdf/39669573.pdf (accessed on 20 April 2021).
21. Bodurov, P.; Genchev, V. Industrial tests with innovative energy saving grinding bodies. *J. Multidiscip. Eng. Sci. Technol. JMEST* **2017**, *4*, 6498–6503.
22. Penchev, T.; Bodurov, P. Comparative Analysis of the Parameters of Spherical and Relo Body Balls for Drum Mills. In Proceedings of the International Conference on Mining, Material and Metallurgical Engineering, Prague, Czech Republic, 11–12 August 2014. Paper No. 144.
23. Coello Velázquez, A.L.; Menéndez-Aguado, J.M.; Brown, R.L. Grindability of lateritic nickel ores in Cuba. *Powder Technol.* **2008**, *182*, 113–115. [CrossRef]
24. Genchev, V.; Bodurov, P.; Kolev, N.; Leroux, D.P. Assessing the Response of Tumbling Mills to the Replacement of Balls by Relo Grinding Media (RGM)—Part 1. Comparative Bench-Scale Experiments and Demonstration Full-Scale Test. In Proceedings of the 52nd Annual Canadian Mineral Processors Operators Conference, Ottawa, ON, Canada, 21–23 January 2020.
25. McIvor, R.E. Why do we need such a high recirculating load on our ball mill? *Metcomtech Grinding Bulletin Issue 4, February 2013*. Available online: https://www.metcomtech.com/grindingbulletin4.php (accessed on 20 April 2021).
26. McIvor, R.E. Plant performance improvements using grinding circuit Classification system efficiency. *Min. Eng.* **2014**, *66*, 67–71.
27. Morrell, S. A new autogenous and semi-autogenous mill model for scale-up, design and optimisation. *Miner. Eng.* **2004**, *17*, 437–445. [CrossRef]
28. Yu, P. A Generic Dynamic Model Structure for Tumbling Mills. Ph.D. Thesis, The University of Queensland, Queensland, Australia, 2014.

Article

Particle Size Distribution Models for Metallurgical Coke Grinding Products

Laura Colorado-Arango [1], Juan M. Menéndez-Aguado [2,*] and Adriana Osorio-Correa [1]

[1] Departamento de Ingeniería Química, Universidad de Antioquia, Medellín 050010, Colombia; laura.coloradoa@udea.edu.co (L.C.-A.); adriana.osorio@udea.edu.co (A.O.-C.)
[2] Escuela Politécnica de Mieres, Universidad de Oviedo, 33600 Oviedo, Spain
* Correspondence: maguado@uniovi.es; Tel.: +34-985458033

Abstract: Six different particle size distribution (Gates–Gaudin–Schuhmann (GGS), Rosin–Rammler (RR), Lognormal, Normal, Gamma, and Swebrec) models were compared under different metallurgical coke grinding conditions (ball size and grinding time). Adjusted R^2, Akaike information criterion (AIC), and the root mean of square error (RMSE) were employed as comparison criteria. Swebrec and RR presented superior comparison criteria with the higher goodness-of-fit and the lower AIC and RMSE, containing the minimum variance values among data. The worst model fitting was GGS, with the poorest comparison criteria and a wider results variation. The undulation Swebrec parameter was ball size and grinding time-dependent, considering greater b values ($b > 3$) at longer grinding times. The RR α parameter does not exhibit a defined tendency related to grinding conditions, while the k parameter presents smaller values at longer grinding times. Both models depend on metallurgical coke grinding conditions and are hence an indication of the grinding behaviour. Finally, oversize and ultrafine particles are found with ball sizes of 4.0 cm according to grinding time. The ball size of 2.54 cm shows slight changes in particle median diameter over time, while 3.0 cm ball size requires more grinding time to reduce metallurgical coke particles.

Keywords: particle size distribution; metallurgical coke; comminution

Citation: Colorado-Arango, L.; Menéndez-Aguado, J.M.; Osorio-Correa, A. Particle Size Distribution Models for Metallurgical Coke Grinding Products. *Metals* **2021**, *11*, 1288. https://doi.org/10.3390/met11081288

Academic Editors: Jürgen Eckert and Luis Norberto López De Lacalle

Received: 30 June 2021
Accepted: 13 August 2021
Published: 16 August 2021

Publisher's Note: MDPI stays neutral with regard to jurisdictional claims in published maps and institutional affiliations.

Copyright: © 2021 by the authors. Licensee MDPI, Basel, Switzerland. This article is an open access article distributed under the terms and conditions of the Creative Commons Attribution (CC BY) license (https://creativecommons.org/licenses/by/4.0/).

1. Introduction

Metallurgical coke is a crucial raw material in the iron and steelmaking industry and is considered a critical raw material in the EU due to its high consumption volume and the strong EU import dependence [1–3]. Heat supplier, reducing agent, adequate permeability, and burden mechanical support are the features that render it a fundamental material for blast furnaces that perform metallurgical processes such as cast iron, ferroalloy, lead, and zinc production, and in kilns for lime and magnesium production [4,5]. According to particle size, metallurgical coke is used at different process stages. Coke ranging between 24–40 mm is the main form for blast furnaces; this so-called nut coke is added in ironmaking with ferrous and flux mineral layers from 6 to 24 mm, and coke breeze is considered the energy source for sintering or pelletising with particle size less than 6 mm [6].

Suitable coke selection enhances the steel production line, saves coke utilisation, minimises dust generation, reduces the significant amount of greenhouse gases discharged into the atmosphere, namely, CO_2, SO_2, and NO_x, and optimises energy usage [7–9]. Chemical composition, mechanical strength, thermal resistance, and particle size are the most significant parameters for selecting metallurgical coke [6,10]. However, the coke particle size and shape play an essential role in blast furnace and sinter plants. Coke mean particle size determines the fluid flow resistance, the upward gases and downwards metal liquids passing efficiency, and the iron production rate. The coke bed formation and permeability are also strongly related to particle size and combustion behaviour in the sintering process.

Despite the abovementioned importance of coke particle size distribution (PSD), particle size control has not been studied enough. Many fine recycled particles from chipping in crushing processes or waste in the coke oven cause an overproduction of particulate matter and uncontrolled coke size distributions [11–22]. Poor coke without quality classification creates disturbances in the sinter plant and the blast furnace operation, producing excess dust, heat losses, inefficient reaction rates, and fluid flow obstruction. This situation has driven environmental regulation to propose eliminating or partially substituting metallurgical coke in sinter production [8,13]. Various studies [14–23] have evaluated the effect of defined ranges of coke particle size in steelmaking performance as a means of process optimisation. The thickness of combustion zone, flame front, chemical reactions kinetics, and iron-bearing phase formation (hematite, magnetite, and gangue) are broadly affected by coke PSD in sinter and blast furnace plants [15–17].

Modelling the metallurgical coke PSD allows quantitatively assessing the breakage behaviour. Several benefits in the iron and steel processes are obtained when the PSD is adequately characterised, with effective diameters (D_{50}, D_{80}) measured, and the effect of its variation on the processes is known. Many models have been developed to predict and describe the PSD of granulated materials. Perfect et al. [24] tested three distribution functions based on two parameters for different fertilisers. Lognormal, Rosin–Rammler and Gate–Gaudin–Schuhmann were fitted by nonlinear regression analysis. According to the goodness-of-fit of R^2, they concluded the Rosin–Rammler is the more accurate model to describe material fertiliser. Botula et al. [25] evaluated ten PSD models in soils of the humid tropics. The findings demonstrated that the three and four-parameters Fredlund and three-parameter Weibull and four-parameter Anderson presented an excellent fitting correlation to soils. Bu et al. [26] characterised the coal grinding process (wet and dry ways) using PSD models, namely, GGS, Gaudin–Meloy, RR, modified RR, and Swebrec. They found that the RR and Swebrec showed outstanding fitting performance.

The current paper compares GGS, RR, Gamma, Normal, Lognormal, and Swebrec distributions at different metallurgical coke grinding conditions to select the best fitting models and represent the metallurgical coke PSD. Finally, the association of PSD model parameters with the grinding process was analysed for the best two models.

2. Materials and Methods

A metallurgical coke sample from Boyacá (Colombia) was used in the grinding process. The original sample was crushed in a roll mill (Denver Equipment Co., Denver, CO, USA) and sieved 100% under six mesh (3.35 mm). Product PSD is depicted in Figure 1, and the elemental and proximate analyses are shown in Table 1. A dataset of 144 PSD was collected from grinding under different dry conditions. Grinding tests were carried out in a laboratory steel ball mill with 0.20 m in diameter and 0.20 m long. Three ball sizes (2.54, 3.00, and 4.00 cm) and eight grinding times (0.5, 1, 2, 3, 4, 5, 6, and 10 min) were used to evaluate the product PSD. The operational mill conditions remained fixed: the fraction of critical speed (φ_c) was 0.75; the ball filling fraction (J) was 0.3; the fraction of powder bed (f_c) and void filling (U) were 0.12 and 1, respectively.

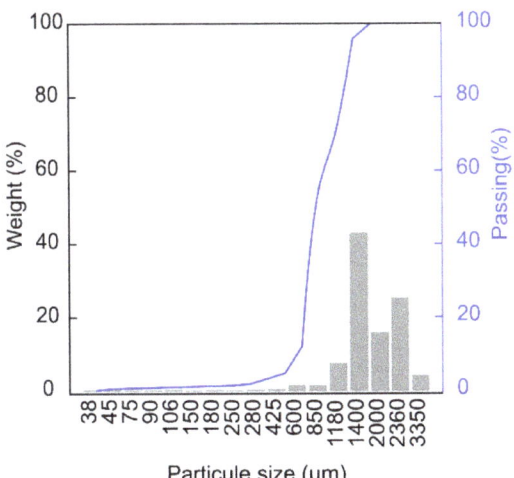

Figure 1. Metallurgical coke PSD in the feed to the ball mill.

Table 1. Metallurgical coke composition analysis (%).

Sample	C	H	O			N		S
Metallurgical Coke	82.55	0.79	14.46			1.3		0.9
Fixed Carbon	Volatiles	Moisture	Ash					
			SiO_2	Fe_2O_3	Al_2O_3	CaO	Others	Σ
76.87	0.91	0.047	28.90	26.73	10.62	10.28	23.45	100

The six PSD models assessed are shown in Table 2. Gate–Gaudin–Schuhmann [27] and Rosin–Rammler [28] models have been the most popular and oldest functions used to describe PSD in granular materials. GGS was developed in the metalliferous mining industry and is described with a size parameter (largest particle size) and a distribution parameter [27]. The RR model was defined to evaluate the coal fragmentation processes but has been broadly used in many industries. The RR size parameter corresponds to 63.21% cumulative undersize, and the shape parameter defines the spread of sizes [28]. Even though these are handy models, the fitting accuracy depends on the material nature and size ranges.

A short description of the most common powder PSD models is presented below. The Gamma distribution [29,30] presents two functional parameters related to the median and standard deviation. Yang et al. [29] compared the PSD prediction between Gamma and other models, obtaining the Gamma distribution as the best fit. Normal and Lognormal are also two-parameter models, using the mean diameter (logarithmic mean if Lognormal) and the standard deviation. According to Buchan [31], the Lognormal is more suitable in describing PSD in soils.

The three-parameter Weibull distribution is defined by fitting, size, and shape parameters. Esmaeelnejad et al. [32] compared different models to describe soil PSD, concluding that the Weibull model was the most accurate for all samples studied. Another three-parameter distribution is the Swebrec distribution, introduced by Ouchterlony [33] to predict PSD by rock blasting and crushing fragmentation. The parameters are the maximum size x_{max}, the size with 50% cumulative undersize x_{50} and the undulation parameter b. In the work of Osorio et al. [34], the Swebrec model was evaluated in the wet grinding process of quartz ore, obtaining an excellent fitting adjustment. Menéndez-Aguado et al. [35] presented the Swebrec distribution to fit sediment particle size distribution with a high correlation between experimental and model data.

Table 2. Particle size distribution models.

Particle Size Models	Cumulative Distribution Function	Independent Variables
Gamma	$F(x_i) = \frac{\Gamma(\beta, x_i/\alpha)}{\Gamma(\beta)}$	α, β
Lognormal	$F(x_i) = \frac{1}{2} + \frac{1}{2} erf\left[\frac{ln x_i - \mu}{\sqrt{2}\sigma}\right]$	GMD [†], σ
Normal	$F(x_i) = \frac{1}{2} + \frac{1}{2} erf\left[\frac{x_i - \mu}{\sqrt{2}\sigma}\right]$	μ, σ
Rosin–Rammler	$F(x_i) = 1 - exp\left[-\left(\frac{x_i}{\lambda}\right)^k\right]$	λ, k
Schuhmann	$F(x_i) = \left(\frac{x_i}{K}\right)^\alpha$	K, α
Swebrec	$F(x_i) = \dfrac{1}{1 + \left[\dfrac{\ln\left(\frac{x_{max}}{x_i}\right)}{\ln\left(\frac{x_{max}}{x_{50}}\right)}\right]^b}$	x_{max}, x_{50}, b

[†]—Geometric Mean Diameter.

In this study, the model comparison was carried out using three statistical indices. The adjusted R^2 (Equation (1)) measures the discrepancy between predicted and observed data [36]. Akaike's information criterion (Equation (2)) examines the model goodness-of-fit imposing penalties for additional fitting parameters [37]. Finally, the mean root of squared error (RMSE) presented in Equation 3 is the residual error, i.e., the information not contained in the model. The criteria selected are widely used in PSD model selection and in assessing model prediction [25,29,32,36]. The adjusted R^2 is a traditional goodness-of-fit measurement, but it is mainly considered in linear models' interpretation. Additionally, to assure the model selection, RMSE and AIC were used. These criteria are more appropriate to measure the goodness-of-fit in nonlinear models [28,38].

$$R^2_{adj} = 1 - \left(\frac{\frac{RSS}{N-P}}{\frac{TSS}{N-1}}\right) \quad (1)$$

where RSS is the residual sum of squares, N is the number of PSD data points, P is the number of independent variables in a particle size distribution model, and TSS is the total sum of squares.

$$AIC = N \cdot \ln\left(\frac{RSS}{N}\right) + 2P \quad (2)$$

$$RMSE = \left(\frac{RSS}{N}\right)^{0.5} \quad (3)$$

A custom Python script was employed in the fitting procedure, which is provided in the Supplementary Material. All models were compared with the experimental PSD data using the least-squares method to find the best fitting parameters, and the model presenting the best values of the three statistics were selected. The least-squares procedure was obtained considering a nonlinear optimisation method, and the residual sum of squares is calculated with the minimisation function established in Equation (4).

$$RSS = \sum_{i=1}^{n}\left(P_{i,measure} - P_{i,predicted}\right)^2 \quad (4)$$

where $P_{i,measured}$ and $P_{i,predicted}$ represent experimental and model cumulative passing material, respectively. Box plots were employed as graphical representation to provide more insights into the different behaviour of PSD models. Finally, metallurgical coke's more stable grinding conditions are defined using a colour map graph about the two best model parameters.

3. Results

3.1. Comparison of PSD Models' Goodness-of-Fit

Figure 2 depicts the Box plots of statistical indices. The model with the better goodness-of-fit was obtained under descriptive statistics (see Table S1) considering the higher adjusted R^2, the smaller RMSE and lower AIC. The adjusted R^2 (Figure 2a and Figure S1) provides values greater than 0.95 in all models, excluding the GGS distribution, which produced adjustments less than 0.8. The Schuhmann distribution data are widely spread out from the mean with the larger standard deviation, as depicted in the Box plot. Lognormal and Normal models explain completely well the experimental PSD with adjustments varying between 0.95–0.99. However, the Lognormal model adjusted slightly better than Normal model due to the great fitting in 4.00 cm grinding media. Gamma, Rosin-Rammler, and Swebrec exhibit values close to 1.0 and relatively narrow dispersion data; therefore, they were considered the models with superior fitting performance, providing an excellent PSD prediction for the material.

Akaike's information criteria (AIC) Box plot is shown in Figure 2b. It was used to compare the model quality fit and identify the better fitting model, for an increment in goodness-of-fit requires lower AIC values. The AIC results were consistent with R^2 and RMSE estimations, achieving minimum values in Gamma, Rosin–Rammler, and Swebrec distributions. However, the Swebrec model presented the least standard deviation. Lognormal and GGS models depicted poor fits with large mean and standard deviation values about AIC estimator.

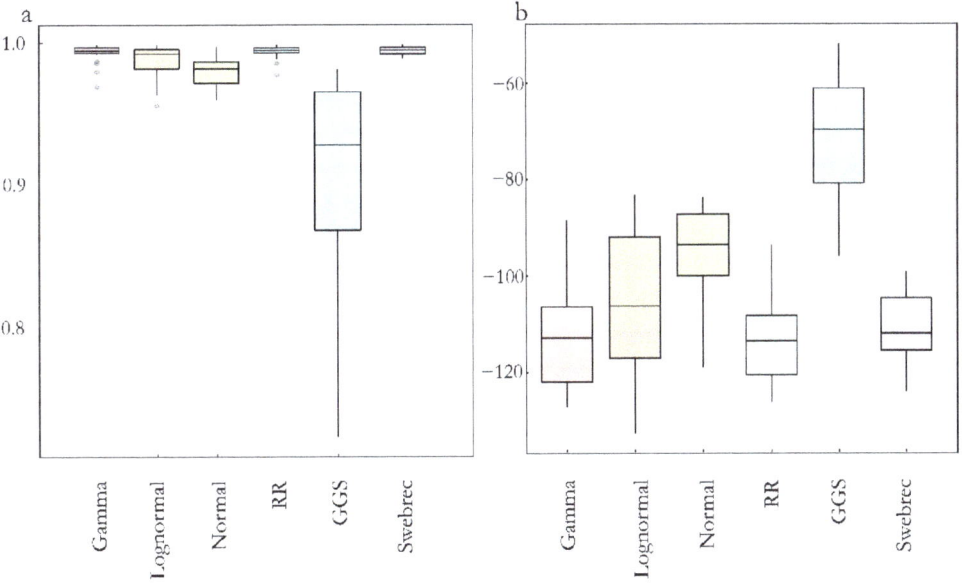

Figure 2. Box plots to compare the particle size models: (**a**) Adjusted R^2 and (**b**) Akaike's information criterion.

Figure 3 summarises the models' criteria in a normalised bar chart. There are three bars (adjusted R^2, AIC, and RMSE) for each distribution, representing the fitting results. GGS and Normal exhibit the larger RMSE values, while Lognormal function shows great AIC values. In addition, GGS presents the lowest adjusted R^2 and larger AIC (the closer this criterion to one, the smaller its actual value). Gamma, RR and Swebrec distributions illustrate the better value points according to the three selected criteria, indicating the excellent correspondence between model prediction values and observed data. Though

AIC penalised the model with additional parameters, the Swebrec model, which has three independent variables, is among the better-fitting functions.

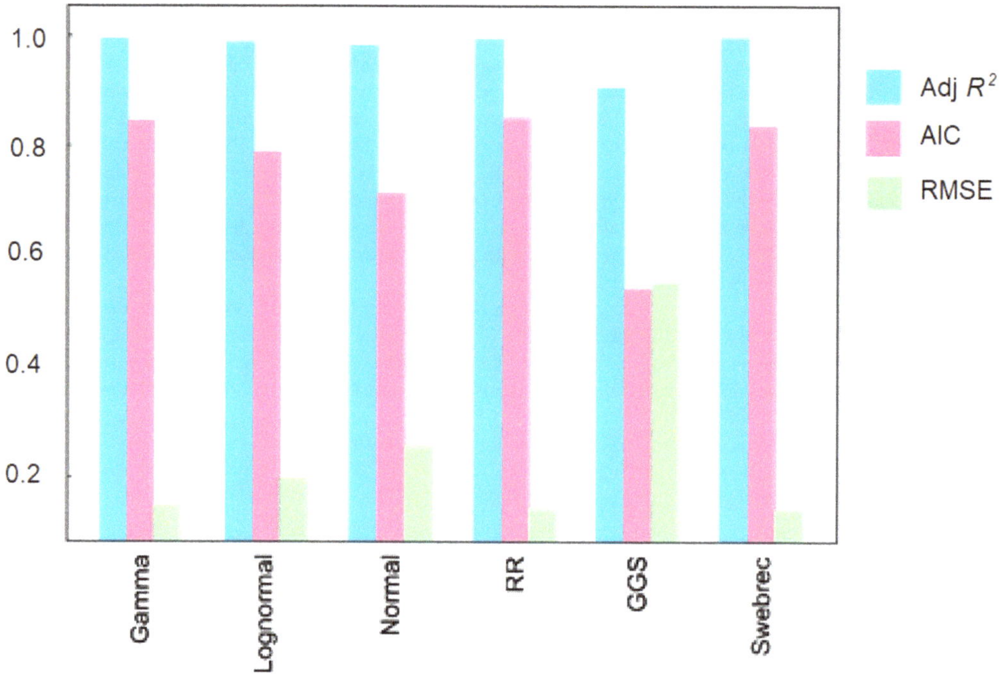

Figure 3. Normalised bar chart of estimator's values for each evaluated model.

3.2. Models' Prediction Ability in Grinding Conditions

Figure 4 shows the PSD obtained (experimental and fitted) after different grinding times in the laboratory ball mill, with 3.00 cm in diameter grinding media balls. The same behaviour was observed in the case of 2.54 cm and 4.00 cm ball diameter (See Supplementary Material, Figures S2 and S3). The GGS model shows a more significant deviation under all the evaluated grinding conditions. After one minute's grinding, the predicted value deviates from the experimental value. This model considers a linear relationship between cumulative fraction and the particle diameter in the log-log scale, where the slope is the α parameter, which does not describe the experimental metallurgical coke grinding product. Normal and Lognormal distributions exhibit, in general, high goodness-of-fit. However, as Normal PSD illustrates, the predicted results decrease at a longer grinding time and smaller grinding media diameter. The Lognormal shows excellent fitting at grinding times over 3 min with all ball sizes, especially with grinding media of 4.00 cm. The curve fitting performance of Gamma, RR, and Swebrec are highly recommended for metallurgical coke grinding products in all evaluated scenarios with an adjusted R^2 range between 0.98–1.0. Model parameter prediction ability for the three abovementioned models were suited correctly with the ball size and grinding time values studied.

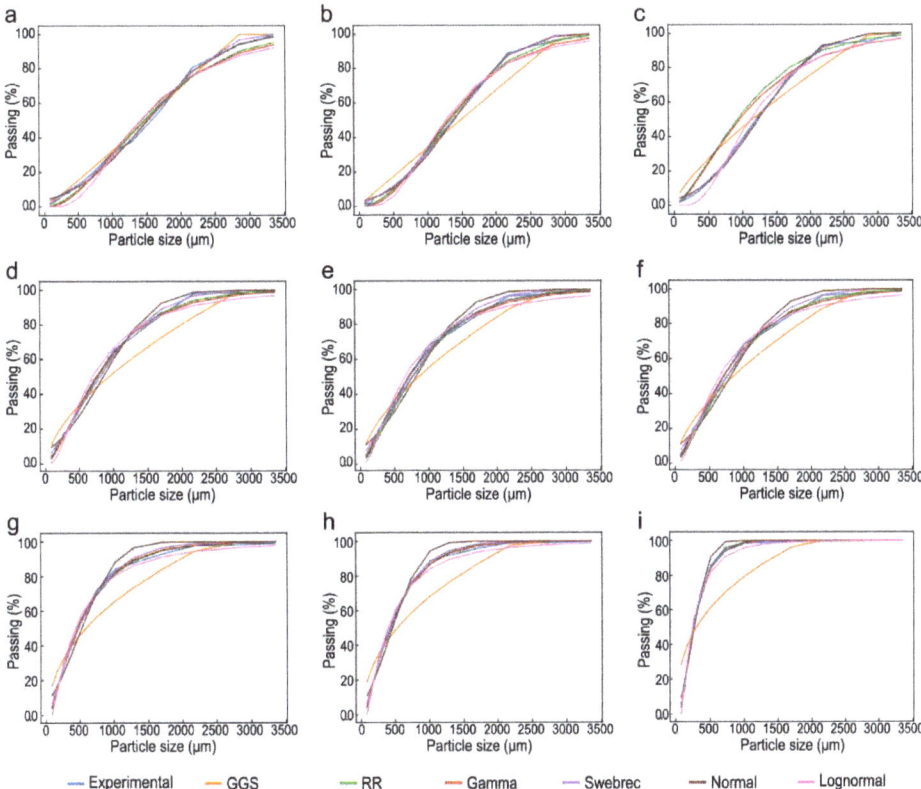

Figure 4. Cumulative distribution functions with 3.0 cm ball size at different grinding times: (**a**) 0 min; (**b**) 0.5 min; (**c**) 1 min; (**d**) 2 min; (**e**) 3 min; (**f**) 4 min; (**g**) 5 min; (**h**) 6 min; (**i**) 10 min.

3.3. Distribution Parameters' Assessment

In light of the above results, Swebrec and RR models were selected to assess the distribution parameters under different grinding conditions. Figure 5a depicts a rise in the undulation parameter b as grinding time increases. Comparing the b parameter variation with different grinding media is relatively stable until 3 min, after which superior b values ($b > 3$) are achieved for all ball sizes. Smaller undulation parameters are obtained with 2.54 cm ball diameter and the larger ones with 4.00 cm diameter. This situation can be associated with the ball energy delivery at grinding time under 3 min; the x_{50} (see Figure 5b) remains directly related to the ball size, with the x_{50} at 2.54 cm being smaller than the x_{50} at 4.00 cm. The undulation parameter can indicate a change in fracture behaviour, ranging from normal breakage at a shorter time and smaller grinding media, passing by chipping, and finally, achieving material pulverisation at a longer time and larger ball diameter.

The λ and k RR parameters are illustrated in Figure 6. The particular behaviour of the shape parameter λ (Figure 6a) between 0 to 8 min grinding time is evidenced. The smaller parameter value, linked with the larger fines quantity, is formed using the ball size range 2.54–3.00 cm. A widening is noticed at grinding times shorter than 2 min, with 3.00 and 4.00 cm ball sizes. Scale parameter k (Figure 6b) presents smaller values at longer grinding times. The value increases from 2.54 to 3.00 cm, decreasing afterwards to 4.00 cm ball size. Larger ball sizes (3.00 and 4.00 cm) lead to greater k values' variation with the grinding time.

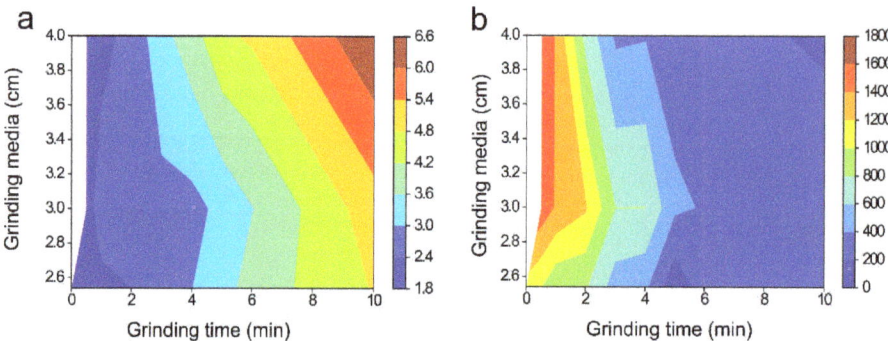

Figure 5. Swebrec parameters' variation under different grinding conditions: (**a**) b, undulation parameter and (**b**) x_{50} parameter.

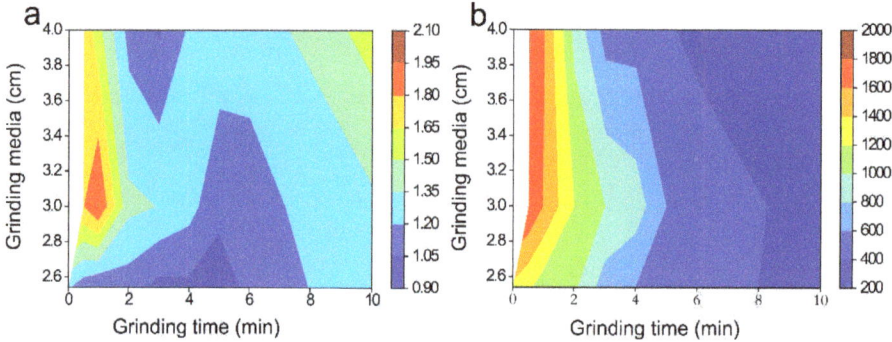

Figure 6. RR parameters' variation under different grinding conditions: (**a**) λ parameter and (**b**) k parameter.

Table 3 shows the x_{50} values measured and predicted by the Swebrec model under different grinding conditions. The median diameter decreases when increasing the grinding time in all cases (as expected, due to the comminution action). The variation is different for each ball diameter; higher grinding kinetics is observed in the case of 4.00 cm ball diameter at grinding times longer than 2 min.

Table 3. Median diameter x_{50} according to Swebrec distribution, at different grinding times.

Grinding Time (min)	Ball Size (cm)					
	2.54		3.00		4.00	
	Measured	Predicted	Measured	Predicted	Measured	Predicted
0.5	1015	1010.03	1520	1523.44	1605	1603.40
1.0	850	840.75	1400	1378.12	1380	1362.85
2.0	830	826.46	1250	1226.03	725	734.81
3.0	512	504.06	800	802.39	360	366.17
4.0	450	448.56	790	800.87	380	386.72
5.0	390	407.5	420	436.00	300	308.40
6.0	280	311.88	380	384.10	216	216.11
10	265	265.44	225	226.40	190	193.22

4. Discussion

Grinding conditions influence the product PSD, and the accuracy of the fitting distribution is highly dependent on the selected model. Statistical indexes used in this study, namely adjusted R^2, AIC, and RMSE, have been widely used to the goodness-of-fit assessing of different PSD and have presented advantages in the models' calibration by least-squares method [25,29,32,35].

Both RR and Swebrec functions show excellent fitting performance for all grinding conditions studied, whereas GGS indicated the poorest yield fitting. These results agree with previous research, which reported better fitting of RR and Swebrec to grinding products and an accurate description of PSD [26,34,39]. Menéndez-Aguado et al. [35] compared different distribution models to sediments and found that the Swebrec model had better performance than Normal, Lognormal, Weibull, and Gamma functions. The undulation Swebrec parameter related to grinding conditions showed dependence on the grinding time and the ball size but only after 4 min for the last mentioned condition. In the case of grinding times less than 4 min, the b parameter seems to be independent of grinding media. As the particle size decreases with the grinding process (Figures 4 and 5), the undulation parameter value increases with high values when chipping action predominates in the fracture process, resulting in a fines overproduction.

Regarding RR parameters, the shape parameter λ (Figure 6a) shows more stability in the central region, which means that the distribution is highly affected by conditions at lower and greater times. Meanwhile, at grinding times of 3–8 min, λ remains between 0.9–1.2, indicating a more wide size interval and a lower slope in the PSD. The scale parameter k (Figure 6b) presents a sensible change in the stability at 4 min grinding time. At grinding time less than 4 min, 3.00 and 4.00 cm ball sizes produce a limited size range with the feature that as ball size increases, the larger fragments do not break, resulting in larger values of k and λ, and indicating an inefficient ball-particle interaction. This situation could indicate that when using 3.00 and 4.00 cm grinding media, the dominant fragmentation mechanism is the chipping abrasion instead of impact breakage.

Although GGS is a popular model used in many sectors, it does not present sound goodness-of-fit in metallurgical coke PSD grinding products under the considered conditions. The GGS predicts distribution with a linear relationship between cumulative weight and particle size in the log-log scale. However, the actual metallurgical coke grinding product shows higher percentage of fines than the model prediction. This result is aligned with previously reported results, which found that the RR gets better PSD fitting than the GGS model [24,40,41].

The Gamma distribution offers an excellent approximation to predict metallurgical coke PSD, but the statistical parameters were more lacking than in the Swebrec and RR models. As shown in Figures 2 and 4, the Gamma function has some outliers in the adjusted R^2 due to shorter grinding times.

Under the grinding test conditions, the production of fines accelerates with all grinding media tested after 3–4 min, producing a widening of the PSD. It has been established in previous studies that the grinding efficiency is related to ball size selection [42–44]. Over time, the metallurgical coke breakage in 4.0 cm balls presents significant variation with undergrinding for 0.5–1 min and overgrinding to grinding time from 3–10 min; 4.00 cm grinding media is perhaps too large, creating voids inside the ball charge and generating less normal forces into particles. These results are consistent with Austin et al. [45] and Khumalo et al. [46] results, which established that the larger ball size action is directed to larger particles whereas small grinding media action is preferent on finer particles. Additionally, Austin et al. [45] proposed that the impact force of collision involving large ball sizes gives larger quantities of fines and more catastrophic fracture behaviour. Ball sizes of 2.54 and 3.00 cm evidence small sudden changes in median diameters. However, lower ball size produces lesser x_{50} sizes, considering balls of 3.00 cm show more grinding to achieve a defined particle size than balls of 2.54 cm. The relationship between ball-particle size for breaking metallurgical coke improves with small ball sizes, possibly due to increased collision frequency.

The assessment of the coke grinding product comparing different particle size functions was carried out. The Swebrec distribution function presented outstanding fitting conditions in grinding products compared to traditional distributions. It is interesting to note that Swebrec parameters, such as GGS and RR models, are related to the fine particles' produced quantities. Therefore, in agreement with other authors [26,34,35], Swebrec function's employment could pose a good alternative to evaluate and control PSD in grinding processes and small particles. Additionally, the metallurgical coke particle range addressed in this study is considered a critical point in steelmaking, especially in sinter plants. The superior sinter properties have been obtained with 3.35 mm undersize [15–17]. As mentioned above, coke particle size influence the sinter porosity, microstructural phases and thermal properties of the sinter bed. The metallurgical coke PSD evaluation in the range between 3.35 mm to 0.212 mm is consistent with Umadevi et al. [15], which found that the use of this size range increases the calcium ferrite phase and decreases the number of the bigger pore size, thus decreasing coke quality and the coke strength index. On the other hand, Dabbagh et al. [16] evaluated the coke PSD effect on the maximum temperature of the sinter bed and the flame front velocity, finding that the particles ranging from 3.35 mm to 0.212 mm increase the heat production and favour the diffusive processes of the sinter bed.

5. Conclusions

Several PSD models were evaluated on metallurgical coke grinding products using adjusted R^2, root means of square error (RMSE) and Akaike's information criterion (AIC) as statistical indices. Variety of grinding media size and grinding time were employed to investigate the goodness-of-fit to PSD models based on different grinding conditions.

- The two better-fitting models, considering grinding media size and grinding time variations, were Swebrec and RR distribution models, presenting superior goodness-of fit-to all defined behaviour conditions.
- The Swebrec distribution undulation parameter b showed larger values to all ball sizes after grinding 4 min with high data at 4.00 cm ball size.
- The Rosin–Rammler λ parameter does not show a defined tendency with grinding conditions. However, with a 2.54 cm ball size, a less spread-out PSD is obtained, ranging λ values between 0.9–1.20. k parameter values are clearly defined with lesser sizes and greater ball size and grinding times.
- The Swebrec and RR models predicted well the metallurgical coke grinding product PSD. Regarding the Swebrec distribution function, it presented excellent fitting conditions in grinding products compared to traditional distributions.
- Oversized and ultrafine particles were found with 4.00 cm ball size depending on the grinding time. The 2.54 cm ball size results showed a slight variation of particle median diameter with time, while 3.00 cm ball size required more grinding time to reduce metallurgical coke particle size.
- The PSD model goodness-of-fit strongly depended on metallurgical coke grinding conditions.

Supplementary Materials: The following are available online at https://www.mdpi.com/article/10.3390/met11081288/s1, Figure S1: Box plot to compare the particle size models base on adjusted R2 criterion, excluding Gates Gaudin Schuhmann distribution, Figure S2: Cumulative distribution functions with 2.54 cm ball size, at different grinding times, Figure S3: Cumulative distribution functions with 4.0 cm ball size, at different grinding times, Table S1: Statistical descriptors for three criteria, Table S2: Custom script to determine fitting parameter by least square method.

Author Contributions: Conceptualisation, L.C.-A., J.M.M.-A. and A.O.-C.; methodology, L.C.-A., J.M.M.-A. and A.O.-C.; software, L.C.-A.; validation, L.C.-A., J.M.M.-A. and A.O.-C.; formal analysis, J.M.M.-A. and A.O.-C.; investigation, L.C.-A.; resources, A.O.-C.; writing—original draft preparation, L.C.-A., J.M.M.-A. and A.O.-C.; writing—review and editing, L.C.-A., J.M.M.-A. and A.O.-C.; visualisation, L.C.-A.; supervision, A.O.-C.; project administration, A.O.-C.; funding acquisition, A.O.-C. All authors have read and agreed to the published version of the manuscript.

Funding: This research received no external funding.

Institutional Review Board Statement: Not applicable.

Informed Consent Statement: Not applicable.

Data Availability Statement: Not applicable.

Acknowledgments: The authors are thankful to the Faculty of Engineering of the Universidad de Antioquia, to the Committee for Research Development, and CODI for economic support to conduct this work in the project framework: Study of alternatives for the improvement of specific rate of breakage through the intensification of the grinding process in ball mill PVR2018 21371.

Conflicts of Interest: The authors declare no conflict of interest.

References

1. Xing, X.; Rogers, H.; Zhang, G.; Hockings, K.; Zulli, P.; Ostrovski, O. Changes in Pore Structure of Metallurgical Cokes under Blast Furnace Conditions. *Energy Fuels* **2015**, *30*, 161–170. [CrossRef]
2. Babich, A.; Senk, D. Coke in the iron and steel industry. In *New Trends in Coal Conversion*; Woodhead Publishing: Sawston, UK, 2018; pp. 367–404. [CrossRef]
3. European Commission. Study on the EU's List of Critical Raw Materials. Available online: https://ec.europa.eu/commission/presscorner/detail/en/ip_20_1542 (accessed on 30 June 2021).
4. Babich, A.; Senk, D.; Gudenau, H.W. Effect of coke reactivity and nut coke on blast furnace operation. *Ironmak. Steelmak.* **2009**, *36*, 222–229. [CrossRef]
5. Huang, J.; Guo, R.; Tao, L.; Wang, Q.; Liu, Z. Effects of Stefan Flow on Metallurgical Coke Gasification with CO_2. *Energy Fuels* **2020**, *34*, 2936–2944. [CrossRef]
6. Cameron, I.; Sukhram, M.; Lefebvre, K.; Davenport, W. *Blast Furnace Ironmaking: Analysis, Control, and Optimisation*; Elsevier: Amsterdam, The Netherlands, 2019.
7. Mousa, E.; Wang, C.; Riesbeck, J.; Larsson, M. Biomass applications in iron and steel industry: An overview of challenges and opportunities. *Renew. Sustain. Energy Rev.* **2016**, *65*, 1247–1266. [CrossRef]
8. Jha, G.; Soren, S.; Mehta, K.D. Partial substitution of coke breeze with biomass and charcoal in metallurgical sintering. *Fuel* **2020**, *278*, 118350. [CrossRef]
9. Fröhlichová, M.; Legemza, J.; Findorák, R.; Maslejová, A. Biomass as a Source of Energy in Iron Ore Agglomerate Production Process. *Arch. Met. Mater.* **2014**, *59*, 815–820. [CrossRef]
10. Geerdes, M.; Chaigneau, R.; Kurunov, I.; Lingiardi, O.; Ricketts, J. *Modern Blast Furnace Ironmaking: An Introduction*; IOS Press: Amsterdam, The Netherlands, 2015.
11. Shin, S.-M.; Jung, S.-M. Gasification Effect of Metallurgical Coke with CO2 and H2O on the Porosity and Macrostrength in the Temperature Range of 1100 to 1500 °C. *Energy Fuels* **2015**, *29*, 6849–6857. [CrossRef]
12. Weitkamp, E.A.; Lipsky, E.M.; Pancras, P.J.; Ondov, J.M.; Polidori, A.; Turpin, B.J.; Robinson, A.L. Fine particle emission profile for a large coke production facility based on highly time-resolved fence line measurements. *Atmos. Environ.* **2005**, *39*, 6719–6733. [CrossRef]
13. Al-Hamamre, Z.; Saidan, M.; Hararah, M.; Rawajfeh, K.; Alkhasawneh, H.E.; Al-Shannag, M. Wastes and biomass materials as sustainable-renewable energy resources for Jordan. *Renew. Sustain. Energy Rev.* **2017**, *67*, 295–314. [CrossRef]
14. Mohamed, F.; El-Hussiny, N.; Shalabi, M. Granulation of coke breeze fine for using in the sintering process. *Sci. Sinter.* **2010**, *42*, 193–202. [CrossRef]
15. Umadevi, T.; Deodhar, A.V.; Kumar, S.; Prasad, C.S.G.; Ranjan, M. Influence of coke breeze particle size on quality of sinter. *Ironmak. Steelmak.* **2008**, *35*, 567–574. [CrossRef]
16. Dabbagh, A.; Moghadam, A.H.; Naderi, S.; Hamdi, M. A study on the effect of coke particle size on the thermal profile of the sinters produced in Esfahan Steel Company (ESCO). *S. Afr. Inst. Min. Metall.* **2013**, *113*, 941–945.
17. Tobu, Y.; Nakano, M.; Nakagawa, T.; Nagasaka, T. Effect of Granule Structure on the Combustion Behavior of Coke Breeze for Iron Ore Sintering. *ISIJ Int.* **2013**, *53*, 1594–1598. [CrossRef]
18. Maeda, T.; Kikuchi, R.; Ohno, K.-I.; Shimizu, M.; Kunitomo, K. Effect of Particle Size of Iron Ore and Coke on Granulation Property of Quasi-Particle. *ISIJ Int.* **2013**, *53*, 1503–1509. [CrossRef]
19. Niesler, M.; Stecko, J.; Blacha, L.; Oleksiak, B. Applications of fine grained coke breeze fractions in the process of iron ore sintering. *Metalurgija* **2014**, *53*, 37–39.

20. Mingshun, Z.; Han, S.; Wang, L.; Jiang, X.; Xu, L.; Zhai, L.; Liu, J.; Zhang, H.; Qin, X.; Shen, F.; et al. Effect of Size Distribution of Coke Breeze on Sintering Performance. *Steel Res. Int.* **2015**, *86*, 1242–1251. [CrossRef]
21. Chung, J.K.; Lee, S.M.; Shin, M.S. Effect of Coke Size on Reducing Agent Ratio (RAR) in Blast Furnace. *ISIJ Int.* **2018**, *58*, 2228–2235. [CrossRef]
22. Xiong, L.; Peng, Z.; Gu, F.; Ye, L.; Wang, L.; Rao, M.; Zhang, Y.; Li, G.; Jiang, T. Combustion behavior of granulated coke breeze in iron ore sintering. *Powder Technol.* **2018**, *340*, 131–138. [CrossRef]
23. Ma, H.; Pan, W.; Liu, L.; Zhang, Z.; Wang, C. Effects of Particle Size of Coke on Iron Ore Sintering Process. *Metal. Sisak Zagreb* **2019**, 649–656. [CrossRef]
24. Perfect, E.; Xu, Q.; Terry, D.L. Improved Parameterization of Fertilizer Particle Size Distribution. *J. AOAC Int.* **1998**, *81*, 935–942. [CrossRef]
25. Botula, Y.-D.; Cornelis, W.M.; Baert, G.; Mafuka, P.; Van Ranst, E. Particle size distribution models for soils of the humid tropics. *J. Soils Sediments* **2013**, *13*, 686–698. [CrossRef]
26. Bu, X.; Chen, Y.; Ma, G.; Sun, Y.; Ni, C.; Xie, G. Wet and dry grinding of coal in a laboratory-scale ball mill: Particle-size distributions. *Powder Technol.* **2019**, *359*, 305–313. [CrossRef]
27. Schuhmann, R., Jr. Energy input and size distribution in comminution. *Trans. SME/AIME* **1960**, *17*, 22–25.
28. Rosin, P.; Rammler, E. Laws governing the fineness of powdered coal. *J. Inst. Fuel* **1933**, *7*, 29–36. [CrossRef]
29. Yang, X.; Lee, J.; Barker, D.E.; Wang, X.; Zhang, Y. Comparison of six particle size distribution models on the goodness-of-fit to particulate matter sampled from animal buildings. *J. Air Waste Manag. Assoc.* **2012**, *62*, 725–735. [CrossRef]
30. Pinho, H.J.O.; Mateus, D.M.R.; Alves, S.S. Probability density functions for bubble size distribution in air–water systems in stirred tanks. *Chem. Eng. Commun.* **2018**, *205*, 1105–1118. [CrossRef]
31. Buchan, G.D. Applicability of the Simple Lognormal Model to Particle-Size Distribution in Soils. *Soil Sci.* **1989**, *147*, 155–161. [CrossRef]
32. Esmaeelnejad, L.; Siavashi, F.; Seyedmohammadi, J.; Shabanpour, M. The best mathematical models describing particle size distribution of soils. *Model. Earth Syst. Environ.* **2016**, *2*, 1–11. [CrossRef]
33. Ouchterlony, F. The Swebrec© function: Linking fragmentation by blasting and crushing. *Trans. Inst. Min. Metall. Sect. A Min. Technol.* **2005**, *114*, 29–44. [CrossRef]
34. Osório, A.; Menendez-Aguado, J.M.; Bustamante, O.; Restrepo, G. Fine grinding size distribution analysis using the Swrebec function. *Powder Technol.* **2014**, *258*, 206–208. [CrossRef]
35. Menéndez-Aguado, J.M.M.; Carpio, E.P.; Sierra, C. Particle size distribution fitting of surface detrital sediment using the Swrebec function. *J. Soils Sediments* **2015**, *15*, 2004–2011. [CrossRef]
36. Shangguan, W.; Dai, Y.; García-Gutiérrez, C.; Yuan, H. Particle-size distribution models for the conversion of Chinese data to FAO/USDA system. *Sci. World J.* **2014**, *2014*, 109310. [CrossRef]
37. Akaike, H.; Parzen, E.; Tanabe, K.; Kitagawa, G. Information theory and an extension of the maximum likehood principle. *Second. Int. Symp. Inf. Theory* **1973**, *3*, 267–281.
38. Spiess, A.-N.; Neumeyer, N. An evaluation of R2 as an inadequate measure for nonlinear models in pharmacological and biochemical research: A Monte Carlo approach. *BMC Pharmacol.* **2010**, *10*, 6. [CrossRef]
39. Liu, S.; Wang, H.; Wang, H. Effect of grinding time on the particle size distribution characteristics of tuff powder. *Medziagotyra* **2021**, *27*, 205–209. [CrossRef]
40. Harris, C.C. A Mull-Purpose Alyavdin-Rosin-Rammler-Weibull Chart. *Powder Technol.* **1971**, *10027*, 3–6. [CrossRef]
41. Allaire, S.; Parent, L.-E. Size Guide Number and Rosin–Rammler Approaches to describe Particle Size Distribution of Granular Organic-based Fertilisers. *Biosyst. Eng.* **2003**, *86*, 503–509. [CrossRef]
42. Austin, L.; Shoji, K.; Luckie, P. The effect of ball size on mill performance. *Powder Technol.* **1976**, *14*, 71–79. [CrossRef]
43. Bwalya, M.; Moys, M.; Finnie, G.; Mulenga, F. Exploring ball size distribution in coal grinding mills. *Powder Technol.* **2014**, *257*, 68–73. [CrossRef]
44. Bürger, R.; Bustamante, O.; Fulla, M.; Rivera, I. A population balance model of ball wear in grinding mills: An experimental case study. *Miner. Eng.* **2018**, *128*, 288–293. [CrossRef]
45. Austin, L.G.; Klimpel, K.K.; Luckie, P.T. *Process Engineering of Size Reduction: Ball Milling*; Society of Mining of the American Institute of Mining, Metallurgical, and Petroleum Enginireers: Englewood, CO, USA, 1984.
46. Khumalo, N.; Glasser, D.; Hildebrandt, D.; Hausberger, B.; Kauchali, S. The application of the attainable region analysis to comminution. *Chem. Eng. Sci.* **2006**, *61*, 5969–5980. [CrossRef]

Article

Study of Comminution Kinetics in an Electrofragmentation Lab-Scale Device

Angel R. Llera [1], Ana Díaz [1], Francisco J. Pedrayes [2], Juan M. Menéndez-Aguado [1,*] and Manuel G. Melero [2]

[1] Escuela Politécnica de Mieres, University of Oviedo, c/Gonzalo Gutiérrez Quirós, 33600 Mieres, Spain; uo46888@uniovi.es (A.R.L.); diazdana@uniovi.es (A.D.)
[2] Departamento de Ingeniería Eléctrica, Electrónica, de Computadores y Sistemas, University of Oviedo, 33204 Gijón, Spain; pedrayesjoaquin@uniovi.es (F.J.P.); melero@uniovi.es (M.G.M.)
* Correspondence: maguado@uniovi.es; Tel.: +34-985-458-033

Abstract: A significant challenge in mineral raw materials comminution is the improvement of process energy efficiency. Conventional comminution techniques, although globally used, are far from being considered power-efficient. The use of high-voltage electric pulses in comminution is a concept that is worthy of study; despite its lack of industrial-scale validation after several decades of lab-scale research, it seems promising as a pretreatment leading to energy savings. In this article, the Cumulative Kinetic Model methodology is adapted to model the comminution effect in an electrofragmentation device, and study a dunite rock ore. The results show that product particle size distribution (PSD) can be predicted with reasonable accuracy using the proposed model.

Keywords: electrofragmentation; comminution; Marx generator; modeling

1. Introduction

Comminution operations are essential in mineral raw materials industries, and estimations of their share in global energy consumption range from 3 to 5% [1–4], so the improvement of process energy efficiency poses a significant challenge in mineral processing technology. Conventional comminution techniques, although globally used, are far from being considered power-efficient. The use of high-voltage electric pulses (HVEP) in comminution is a concept worth studying; despite its lack of industrial-scale validation after several decades of lab-scale research, it seems promising as a pretreatment leading to energy savings. Moreover, it is probably the only known comminution technology capable of maintaining its efficiency in a zero-gravity environment.

Initial research into HVEP use in comminution started in the mid-20th century to produce rock weakening and selective mineral fragmentation [5,6]. Some studies performed comparisons with conventional technologies on such issues as size reduction capability and energy consumption [7–10], while other studies focused on improving mineral liberation [11–15].

This study proposes a mathematical model to predict product PSD in an HVEP device after one or more electric pulses under specific working conditions. Preliminary tests showed the particular influence of pulse polarity on breakage results, so this effect will also be analyzed.

2. Experimental

2.1. Materials

Samples were supplied by the mineral processing plant at Mina David (Pasek Minerales), located in Landoi (Spain). This is the only dunite producer in Spain; despite the olivine content being too low (20–30%) to classify it as a dunite rock, it keeps this commercial denomination. Along with olivine, it is usually accompanied by orthopyroxene (8–16%), amphibole (14–20%) and chrysotile (0–33%). Moreover, other minerals can appear

in the open pit due to hydrothermal alterations, such as chlorite, serpentinite and clay group minerals. Table 1 shows the X-ray fluorescence (XRF) results. Further characterizations of this ore can be found in [16].

Table 1. XRF ore results (%) (L.O.I. = lost on ignition).

SiO$_2$	Al$_2$O$_3$	Fe$_2$O$_3$	MgO	CaO	K$_2$O	Others	L.O.I.
39.86	3.00	7.62	35.34	1.73	0.07	0.35	11.91

Due to the high Mg content shown above, Pasek Minerales is currently developing an extraction process, aimed at producing high-quality magnesium oxide from dunite fines; any step towards a reduction in the specific energy consumption in the fines production process would be desirable.

To provide comminution characterization, a Bond ball mill standard test was performed on a representative sample, with a result of 11.6 kWh/t at 100 microns.

A sufficient amount of sample was prepared within narrow size intervals via sieving. These fractions can be considered monosizes, and they were tested separately to determine the influence of particle size. The selected intervals were (in microns): 5000/3350; 3350/2000; 2000/1000; 1000/500; 500/125 and 125/0. Table 2 shows the total weights of each monosize after sieving. Aliquots of 500 g were prepared for each monosize using a Jones sample divider (RETSCH, Haan, Germany).

Table 2. Sample weight after preparation.

Monosize (μm)	Weight (kg)
125/0	15.2
500/125	23.12
1000/500	18.34
2000/1000	24.46
3350/2000	12.34
5000/3350	18.30

2.2. Methods

2.2.1. HVEP Test Rig

The test rig (see Figure 1) is based on a Marx pulse generator SGSA 400-20 (HAEFELY, Basel, Switzerland), located at the Electrical Engineering Department facilities in Gijon (University of Oviedo, Spain). The main characteristics of this HVEP test rig are depicted in Table 3.

Table 3. General specifications of the HVEP generator.

Parameter (unit)	Value
Maximum voltage (kV)	400
Maximum energy discharge (kJ)	20
Number of stages	4
Capacity/stage (μF)	1

Figure 2 shows the diagram of a Marx impulse generator. The depicted C and Cs correspond to the test cell and the impulse capacitance, respectively. Rs and Rp are the resistances that define the pulse leading edge time and trailing edge time, respectively. The element SF represents the spark gap that starts the discharge of the impulse capacitance into the test cell, thus generating the requested pulse.

Figure 1. Test rig (1) pulse generator, (2) charge unit, (3) capacitive divider, (4) compensation circuit.

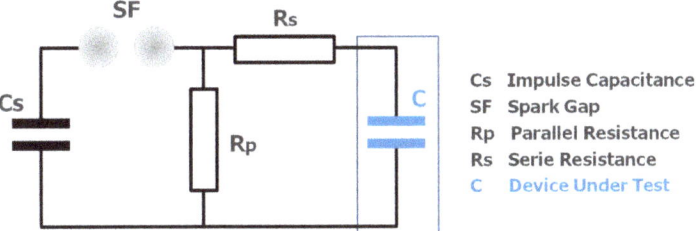

Figure 2. Marx generator diagram.

The pulse generation and measurement process are represented in Figure 3. Firstly, the desired impulse specifications are set in the generator control unit, including the number of work stages, peak impulse value, capacitance charging time and impulse polarity. Secondly, the charging unit raises the voltage to the specified peak value and the charging rectifier converts this to direct current, which is used to charge the generator capacitors. Afterwards, when the capacitors reach the pre-set voltage, the control unit orders the impulse to discharge on the sample within the test cell. Finally, the impulse is registered using a voltage divider in parallel, which permits the signal's digitalization and treatment.

In contrast to the devices used in previous studies [17–20], this test rig has the option of changing impulse polarity. This feature can be achieved by changing the positions of the charge unit diodes (Figure 4), inverting the voltage discharge polarity and thus getting positive or negative discharge impulses on the test sample. Figure 5 shows two examples of no-load impulse curves of different polarities (X axis time in microseconds; Y axis voltage in kV).

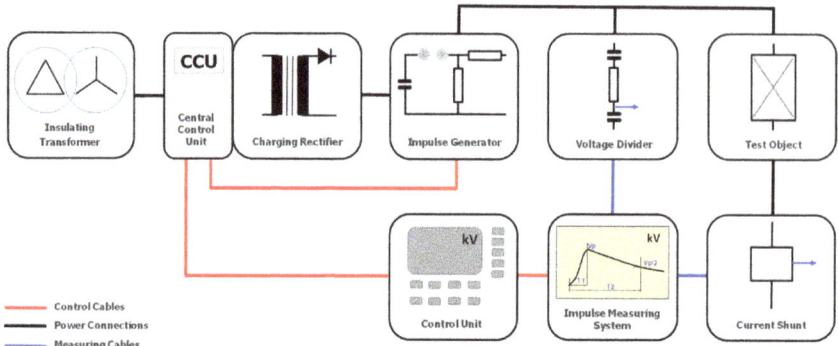

Figure 3. HVEP test rig block diagram.

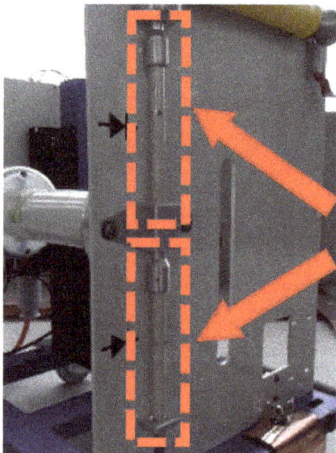

Figure 4. Charge unit diodes.

A relevant parameter in the electrofragmentation tests is the pulse rise time, for this must be short enough to produce a successful fragmentation [21]. Impulse discharge through a mineral sample requires enough voltage to overcome the sample dielectric strength, but the voltage achieved should not surpass the surrounding material's dielectric strength, because, in that case, the discharge would concentrate in the surrounding medium. Additionally, if a medium with higher electric permittivity surrounds the mineral sample, a very uneven distribution of the applied electric field occurs, with a high concentration in the mineral and a much lower concentration in the surrounding medium.

Both effects can be achieved by soaking the mineral sample in distilled water; at a very short pulse rise time, water's dielectric strength and permittivity are higher than rock's [21,22], as shown in Figure 6, which shows that the pulse rise time should be less than 500 ns.

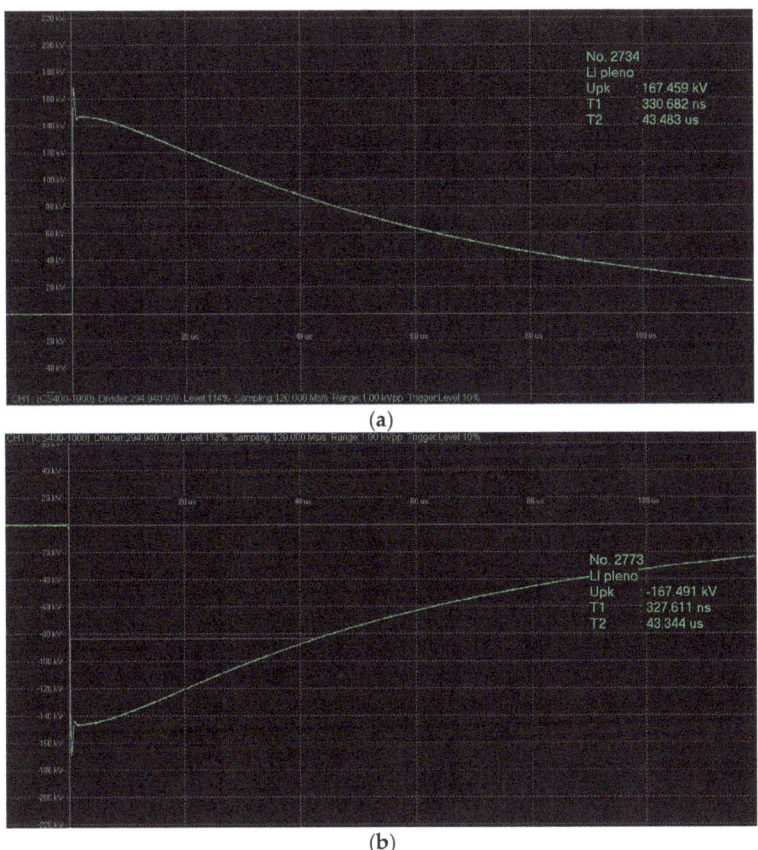

Figure 5. No-load impulses: (**a**) positive polarity; (**b**) negative polarity.

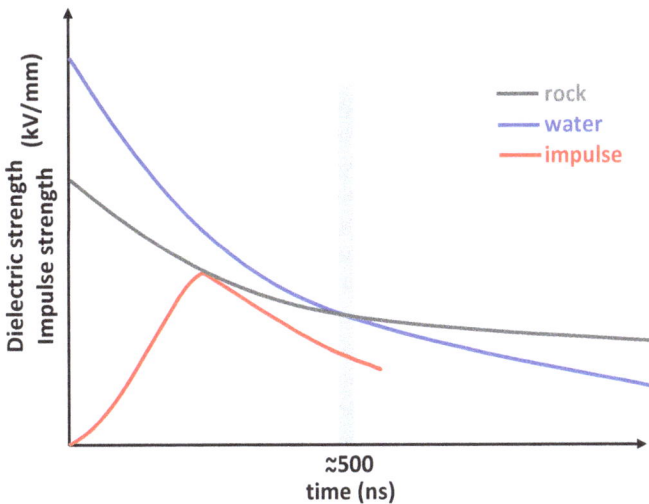

Figure 6. Variation of dielectric strength with the pulse rise time.

With the aim of a more significant reduction in the pulse rise time, we substituted the resistance Rs (Figure 2) for a short-circuit; thus, a pulse rise time around 300 ns can be achieved, with a peak voltage of 150 kV (this value was set in all tests performed), plus an additional value due to overshooting. Under these conditions, the discharge effect will concentrate in the mineral sample; the wave shapes obtained when applying these pulses (both with positive and negative polarity) are shown in Figure 7.

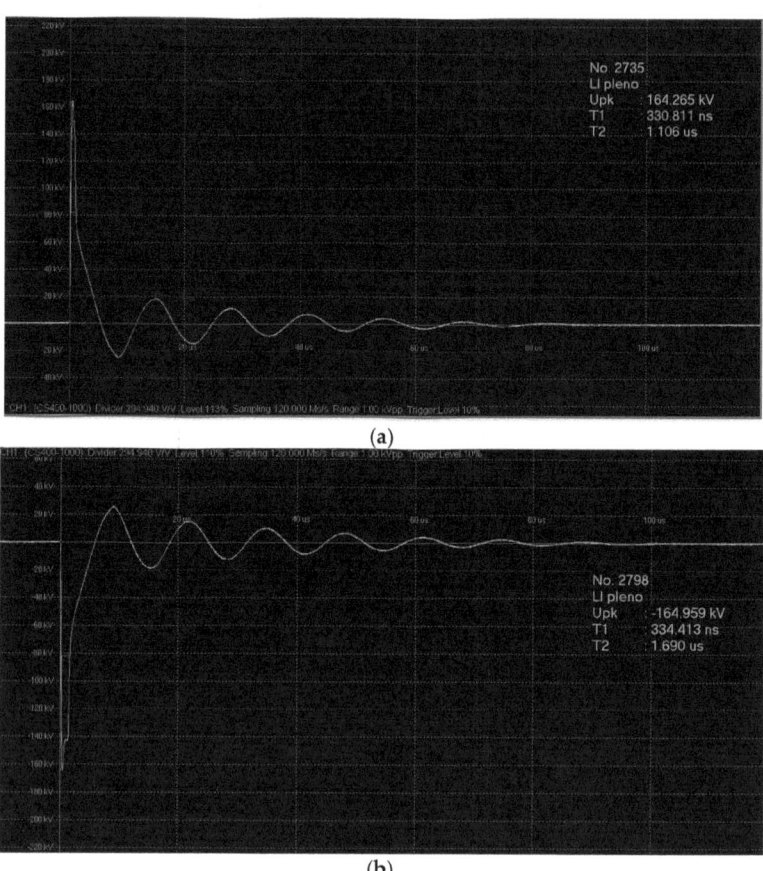

Figure 7. Wave shapes after impulse discharge on dunite sample: (**a**) positive polarity; (**b**) negative polarity.

2.2.2. HVEP Test Cell

In order to correctly apply generated pulses to the mineral sample, a test cell was developed that was to be attached to the Marx pulse generator, following the scheme proposed in [21]. Because the peak voltage values could reach hundreds of kV, the insulator definition, electrode configuration and distances among live elements and grounded elements were critical.

The basis of the test cell was an inox steel vessel acting as the grounded electrode. This vessel has a high-density polyethylene (HDPE) shell inside it that acts as an insulator. The active electrode is also embedded in HDPE and is supported by 3D printed parts that stabilize the whole (Figures 8 and 9), so a flat-tip electrode configuration is defined.

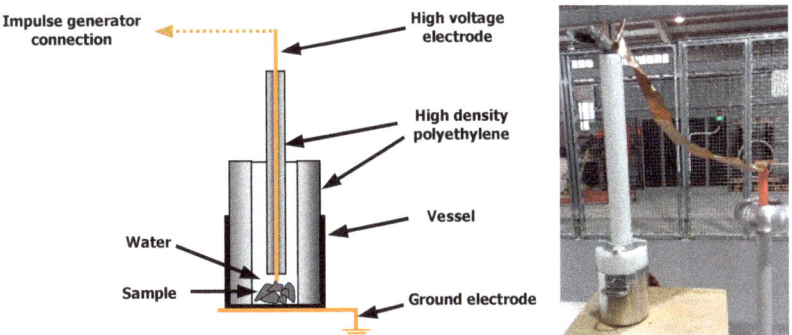

Figure 8. (**Left**): Test cell diagram. (**Right**): Test cell connected to the impulse generator.

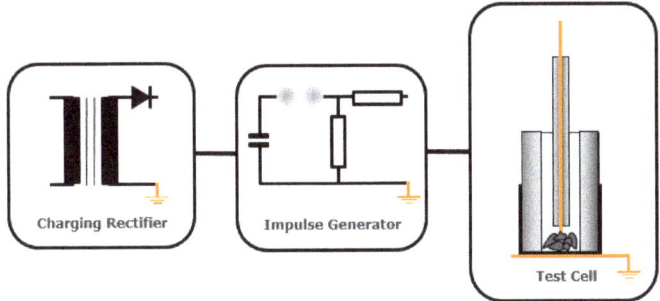

Figure 9. Diagram of the pulse generator and coupled test cell.

The mineral sample and the dielectric liquid are placed at the bottom of the steel vessel, which, in turn, rests on a grounded copper sheet. The active electrode, connected to the pulse generator output, comprises a copper rod that comes into contact with the sample. The HDPE cylindrical pieces guarantee that no electric arcs are formed outside the sample volume. With this electrodes configuration and the expected voltage values, the electrode distance was estimated at 25 mm; this value is in line with the values reported in [11,20–22], within the interval 20–40 mm.

2.2.3. HVEP Test Procedure

The tests were carried out on the pulse generator, applying high-voltage electrical pulses. At each monosize, a total of fourteen tests was performed, seven tests with positive polarity and seven more with negative polarity, in order to establish the possible influence of polarity on the degree of fragmentation of the sample. At each polarity, four samples were tested with one, two, three and four pulses, respectively. The three remaining samples were tested using five pulses to determine the test's repeatability on the final PSD.

After each test, the collected sample was dried to remove the distilled water used as a dielectric medium and sieved to obtain the PSD.

2.2.4. Mathematical Model

A mathematical model that describes the effect of electrofragmentation on PSD is proposed, based on an adaptation of the Cumulative Kinetic Model [23,24] into a discontinuous process, as expressed in Equation (1).

$$W_{(x,i)} = W_{(x,f)} \cdot e^{-k \cdot i} \tag{1}$$

wherein:

$W_{(x,i)}$ is the cumulative oversize of size class x after i pulses;
$W_{(x,f)}$ is the cumulative oversize of size class x in the feed;
k is the breakage rate parameter.

The relationship between the breakage rate parameter and the particle size is shown in Equation (2):

$$k = a \cdot x^b \qquad (2)$$

where a and b can be determined experimentally. Accordingly, once one has defined the model parameters, the electrofragmentation product PSD after i pulses can be obtained from the feed PSD using Equation (3).

$$W_{(x,i)} = W_{(x,0)} \cdot e^{-a \cdot x^b \cdot i} \qquad (3)$$

The k value is determined for each monosize after taking logarithms at Equation (1):

$$\ln(W_{(x,i)}) = \ln(W_{(x,0)}) - k \cdot i \Rightarrow \ln(W_{(x,i)}) - \ln(W_{(x,0)}) = k \cdot i \qquad (4)$$

Once one has obtained k values for each monosize, an additional linear regression can be performed to calculate a and b, according to Equation (5).

$$\ln(k) = \ln(a) + b \cdot \ln(x) \qquad (5)$$

3. Results and Discussion

Tables S1–S10 show the results of the 70 impulse tests performed on different monosizes, with positive and negative polarity, including the three five-pulse replicas.

Figure 10 compares the PSD values (cumulative oversize) in the monosize 5000/3350 μm case when using different polarities. In the case of no influence of the polarity, values should be randomly spread following the diagonal line. However, in this case, plotted points are located above the diagonal line due to the cumulative oversize value being higher in the case of positive polarity; this means that the comminution effect is higher in the case of negative polarity. This monosize shows the same behavior in the case of one to four pulses, while in the case of five pulses, values almost fit the diagonal, thus meaning that polarity does not influence the PSD after five pulses.

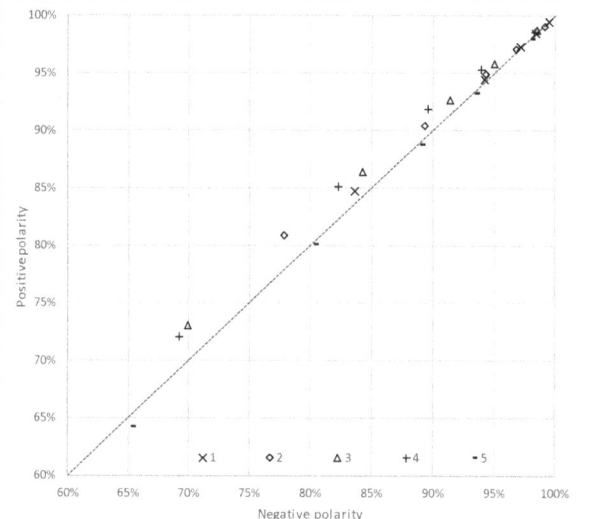

Figure 10. Product PSD after a different number of pulses (feed monosize 5000/3350 μm) and different polarity.

The same analysis was performed with the rest of the monosizes. Figure 11 shows the result in the 3350/2000 µm size interval case, which shows an opposite behaviour from the previous monosize. In this case, the positive polarity seems to produce a more intense comminution effect in the case of one to four pulses, while again, in the case of five pulses, the polarity seems not to influence the PSD. On the other hand, with monosizes 2000/1000 µm, 1000/500 µm, and 500/125 µm (Figures 12–14), the results suggest that the polarity does not influence the comminution effect. From these results, the influence of the polarity cannot be concluded; however, under certain conditions, the results show that a specific polarity could improve the comminution effect in the electrofragmentation device.

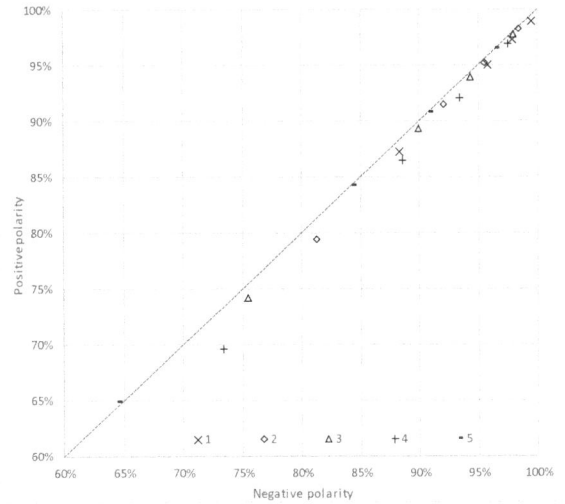

Figure 11. Product PSD after a different number of pulses (feed monosize 3350/2000 µm) and different polarity.

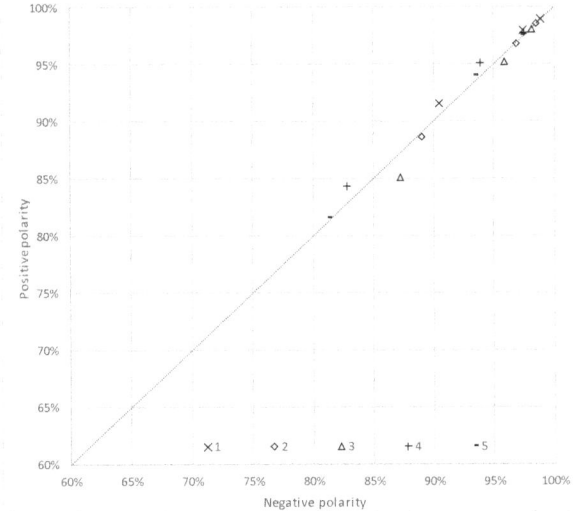

Figure 12. Product PSD after a different number of pulses (feed monosize 2000/1000 µm) and different polarity.

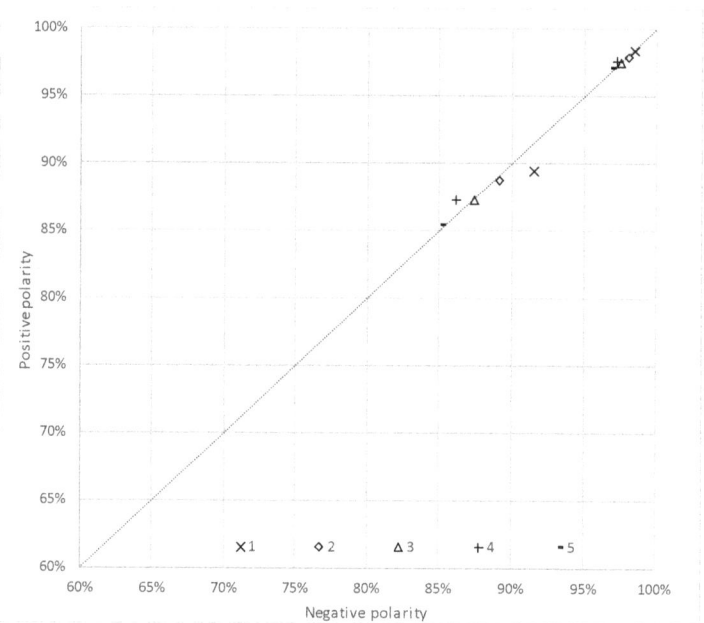

Figure 13. Product PSD after a different number of pulses (feed monosize 1000/500 μm) and different polarity.

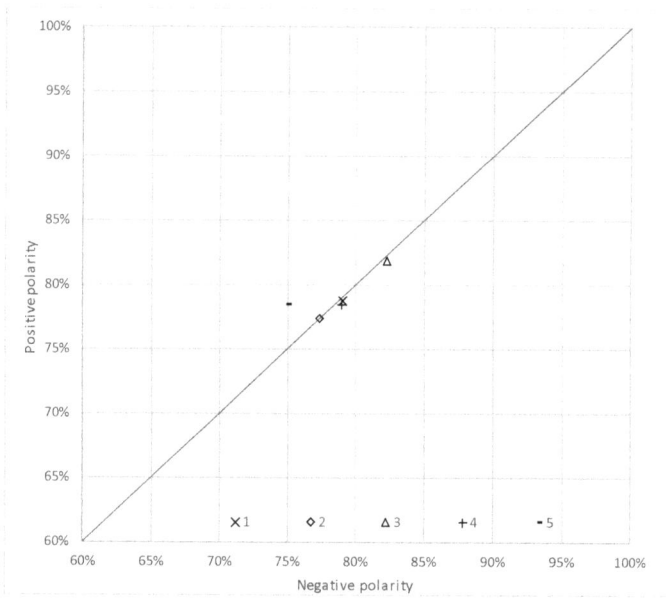

Figure 14. Product PSD after a different number of pulses (feed monosize 500/125 μm) and different polarity.

Regarding the comminution modeling, from data gathered in Tables S1–S10 and Equations (4) and (5), the proposed model parameters can be calculated, again for each polarity. Table 4 shows the results of *a* and *b* parameters and the correlation coefficient value obtained in Equation (5) for linear regression. According to the R^2 values, both polarities show a better fit at coarser monosizes, with very similar values.

Table 4. Model parameter values calculated.

Monosize	Negative Polarity			Positive Polarity		
(μm)	a	b	R^2	a	b	R^2
5000/3350	0.00006	0.85365	0.99730	0.00004	0.90009	0.99940
3350/2000	0.00014	0.78421	0.97140	0.00017	0.77116	0.97210
2000/1000	0.00019	0.63905	0.88830	0.00010	0.72087	0.88710
1000/500	0.00076	0.41263	0.68810	0.00057	0.46036	0.68810

The parameter values shown in Table 4 were calculated by considering replica 1 at five pulses, in order to compare the model's estimated PSD with the remaining replicas. Table 5 gathers the results obtained with both polarities, in the case of the 5000/3350 μm monosize; these results are also plotted in Figure 15. Tables S11–S13 in the Supplementary Material gather the results of the other monosizes.

Table 5. PSD values (modeled and real), feed 5000/3350 μm monosize, five pulses.

Size	Negative Polarity			Positive Polarity		
(μm)	Model	Replica 2	Replica 3	Model	Replica 2	Replica 3
3350	61.54%	65.94%	63.61%	63.24%	61.95%	64.68%
2000	77.29%	80.73%	79.70%	78.62%	79.42%	79.74%
1000	87.03%	89.31%	88.67%	88.17%	88.20%	88.42%
500	92.59%	93.69%	93.12%	93.39%	92.88%	93.04%
125	97.62%	98.09%	97.85%	97.97%	97.81%	97.92%

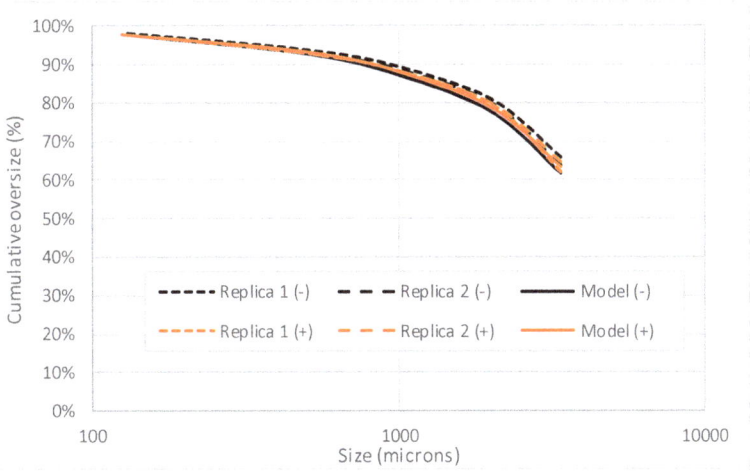

Figure 15. Product PSD after five pulses, monosize 5000/3350 μm.

In order to analyses the results, a first comparison was made between replicas 2 and 3. Subsequently, a second comparison was performed between the modeled PSD values and the average distribution obtained from replicas 2 and 3 (labeled as real). Model deviation had a relative error lower than 2%, which was even lower than 0.5% at finer monosizes. The F-test values are shown in Table 6 for all monosizes and both polarities.

Table 6. F-test values obtained in the comparisons performed.

Monosize (μm)	Negative Polarity		Positive Polarity	
	Among Replicas	Model/Real	Among Replicas	Model/Real
5000/3350	0.9088	0.8724	0.8876	0.9777
3350/2000	0.9974	0.9789	0.9995	0.9561
2000/1000	0.9833	0.9802	0.9943	0.9825
1000/500	0.9928	0.9919	0.9920	0.9836
500/125	0.9891	0.8543	0.9218	0.9868

According to the results shown in Figure 15, in general terms, the proposed model achieves a good fitting of PSD after five pulses, with a slightly better result in the case of positive polarity; this can also be deduced from the F-values shown in Table 6, obtaining a value of 0.9777 in the case of positive polarity, which is higher than the value obtained in the case of negative polarity, 0.8724. Further research must be performed with different ores and pulse conditions to define the influence of pulse polarity.

4. Conclusions

From the results obtained in this research, the following conclusions can be highlighted:

- With a monosize 5000/3500 μm, a negative polarity achieved a better comminution effect, while with the monosize 3350/2000 μm, a positive polarity achieved better performance. In finer monosizes, the polarity effect was not conclusive. Accordingly, the influence of the polarity on the electrofragmentation effect cannot be concluded, and further studies should be performed;
- The proposed model can achieve a good prediction of the electrofragmentation product PSD, after a given number of impulses. The results for both polarities were similar, with a slightly better result in the case of positive polarity.

Supplementary Materials: The following supporting information can be downloaded at: https://www.mdpi.com/article/10.3390/met12030494/s1, Table S1: Results obtained with monosize 5000/3350, negative polarity; Table S2. Results obtained with monosize 5000/3350, positive polarity; Table S3. Results obtained with monosize 3350/2000, negative polarity; Table S4. Results obtained with monosize 3350/2000, positive polarity; Table S5. Results obtained with monosize 2000/1000, negative polarity; Table S6. Results obtained with monosize 2000/1000, positive polarity; Table S7. Results obtained with monosize 1000/500, negative polarity; Table S8. Results obtained with monosize 1000/500, positive polarity; Table S9. Results obtained with monosize 500/125, negative polarity; Table S10. Results obtained with monosize 500/125, positive polarity; Table S11: PSD values (modeled and real), feed 3350/2000 monosize; Table S12: PSD values (modeled and real), feed 2000/1000 monosize; Table S13: PSD values (modeled and real), feed 1000/500 monosize.

Author Contributions: Conceptualization, A.R.L., J.M.M.-A. and M.G.M.; methodology, J.M.M.-A., M.G.M. and F.J.P.; software, A.R.L. and A.D.; validation, F.J.P. and M.G.M.; formal analysis and investigation, A.R.L., F.J.P. and M.G.M.; resources, J.M.M.-A. and M.G.M.; writing—original draft preparation, A.R.L., A.D. and F.J.P.; writing—review and editing, J.M.M.-A. and M.G.M.; supervision, J.M.M.-A. and M.G.M. All authors have read and agreed to the published version of the manuscript.

Funding: This research was partially funded by the Spanish Ministry of Economy and Competitiveness, under project DPI2017-83804-R.

Institutional Review Board Statement: Not applicable.

Informed Consent Statement: Not applicable.

Data Availability Statement: Not applicable.

Conflicts of Interest: The authors declare no conflict of interest.

References

1. Fuerstenau, D.W.; Phatak, P.B.; Kapur, P.C.; Abouzeid, A.-Z.M. Simulation of the grinding of coarse/fine (heterogeneous) systems in a ball mill. *Int. J. Miner. Processing* **2011**, *99*, 32–38. [CrossRef]
2. Schönert, K. Aspects of the physics of breakage relevant to comminution. In Proceedings of the 4th Tewksbury Symposium Fracture, Melbourne, Australia, 12–14 February 1979; pp. 3:1–3:30.
3. Austin, L.G.; Klimpel, R.R. The Theory of Grinding Operations. *Ind. Eng. Chem.* **1964**, *56*, 18–29. [CrossRef]
4. Deniz, V.; Ozdag, H. A new approach to bond grindability and work index: Dynamic elastic parameters. *Miner. Eng.* **2003**, *16*, 211–217. [CrossRef]
5. Usov, A.F.; Tsukerman, V.A. New innovative technologies of processing of mineral raw materials on a basis electric pulse disintegration. In *XXVIII International Mineral Processing Congress*; Science Press: Beijing, China, 2008; pp. 325–328.
6. Shi, F.; Weh, A.; Manlapig, E.; Wang, E. Recent Developments in high voltage electrical comminution research and its potential applications in the mineral industry. In *XXVI Internation Mineral Processing Conference*; Technowrites: New Delhi, India, 2012; pp. 4950–4962.
7. Shi, F.; Zuo, W.; Manlapig, E. Characterisation of pre-weakening effect on ores by high voltage electrical pulses based on single-particle tests. *Miner. Eng.* **2013**, *50–51*, 69–76. [CrossRef]
8. Shi, F.; Zuo, W.; Manlapig, E. Pre-concentration of copper ores by high voltage pulses. Part 2: Opportunities and challenges. *Miner. Eng.* **2015**, *79*, 315–323. [CrossRef]
9. Huang, W.; Shi, F. Improving high voltage pulse selective breakage for ore-preconcentration using a multiple particle treatment method. *Miner. Eng.* **2018**, *128*, 195–201. [CrossRef]
10. Usov, A.F.; Tsukerman, V.A. Innovation processes and device for processing of industrial minerals, mica, diamonds based electric pulse disintegration: Design and testing. In *Abstract of the XXII International Mineral Processing Congress*; South African Institute of Mining & Metallurgy: Cape Town, Africa, 2003.
11. Wang, E.; Shi, F.; Manlapig, E. Mineral liberation by high voltage pulses and conventional comminution with same specific energy levels. *Miner. Eng.* **2012**, *27–28*, 28–36. [CrossRef]
12. Andres, U. Electrical disintegration of rock. *Miner. Processing Extr. Metall. Rev.* **1994**, *14*, 87–110. [CrossRef]
13. Chernet, T. High Voltage Selective Fragmentation for Detailed Mineralogical and Analytical Information, Case Study: Oiva's Gold-Quartz-Dyke in the Lapland Granulite Belt, Laanila, Northern Finland. In *Proceedings of the 10th International Congress for Applied Mineralogy (ICAM)*; Broekmans, M., Ed.; Springer: Berlin/Heidelberg, Germany, 2012.
14. Finkelstein, G.A.; Shuloyakov, A.D. On prospects of electric pulse disintegration from energy balance standpoint. *Miner. Processing Extr. Metall. Rev.* **1996**, *16*, 167–174. [CrossRef]
15. Zuo, W.; Shi, F.; Manlapig, E. Modelling of high voltage pulse breakage of ores. *Miner. Eng.* **2015**, *83*, 168–174. [CrossRef]
16. Baragaño, D.; Forján, R.; Menéndez Aguado, J.M.; Covián Martino, M.; Díaz García, P.; Martínez Rubio, J.; Álvarez Rueda, J.J.; Gallego, J.L.R. Reuse of Dunite Mining Waste and Subproducts for the Stabilization of Metal(oid)s in Polluted Soils. *Minerals* **2019**, *9*, 481. [CrossRef]
17. Touzé, S.; Bru, K.; Ménard, Y.; Weh, A.; Von der Weid, F. Electrical fragmentation applied to the recycling of concrete waste—Effect on aggregate liberation. *Int. J. Miner. Processing* **2017**, *158*, 68–75. [CrossRef]
18. Wang, E.; Shi, F.; Manlapig, E. Pre-weakening of mineral ores by high voltage pulses. *Miner. Eng.* **2011**, *24*, 455–462. [CrossRef]
19. Wang, E.; Shi, F.; Manlapig, E. Factors affecting electrical comminution performance. *Miner. Eng.* **2012**, *34*, 48–54. [CrossRef]
20. Van der Wielen, K.P.; Pascoe, R.; Weh, A.; Wall, F.; Rollinson, G. The influence of equipment settings and rock properties on high voltage breakage. *Miner. Eng.* **2013**, *46–47*, 100–111. [CrossRef]
21. Bluhm, H.; Frey, W.; Giese, H.; Hoppé, P.; Schultheiss, C.; Strässner, R. Application of Pulsed HV Discharges to Material Fragmentation and Recycling. *IEEE Trans. Dielectr. Electr. Insul.* **2000**, *7*, 625–636. [CrossRef]
22. Voigt, M.; Anders, E.; Lehmann, F.; Mezzetti, M.; Will, F. "Electric Impulse Technology—Breaking Rock". In Proceedings of the 22nd European Conference on Power Electronics and Applications EPE'20 ECCE Europe, Lyon, France, 7–11 September 2020.
23. Ersayin, S.; Sönmez, B.; Ergün, L.; Aksani, B.; Erkal, F. Simulation of the Grinding Circuit at Gümüşköy Silver Plant, Turkey. *Trans. Inst. Min. Metall. Sect. C (UK)* **1993**, *102*, C32–C38.
24. Ciribeni, V.; Menéndez-Aguado, J.M.; Bertero, R.; Tello, A.; Avellá, E.; Paez, M.; Coello-Velázquez, A.L. Unveiling the Link between the Third Law of Comminution and the Grinding Kinetics Behaviour of Several Ores. *Metals* **2021**, *11*, 1079. [CrossRef]

MDPI
St. Alban-Anlage 66
4052 Basel
Switzerland
Tel. +41 61 683 77 34
Fax +41 61 302 89 18
www.mdpi.com

Metals Editorial Office
E-mail: metals@mdpi.com
www.mdpi.com/journal/metals